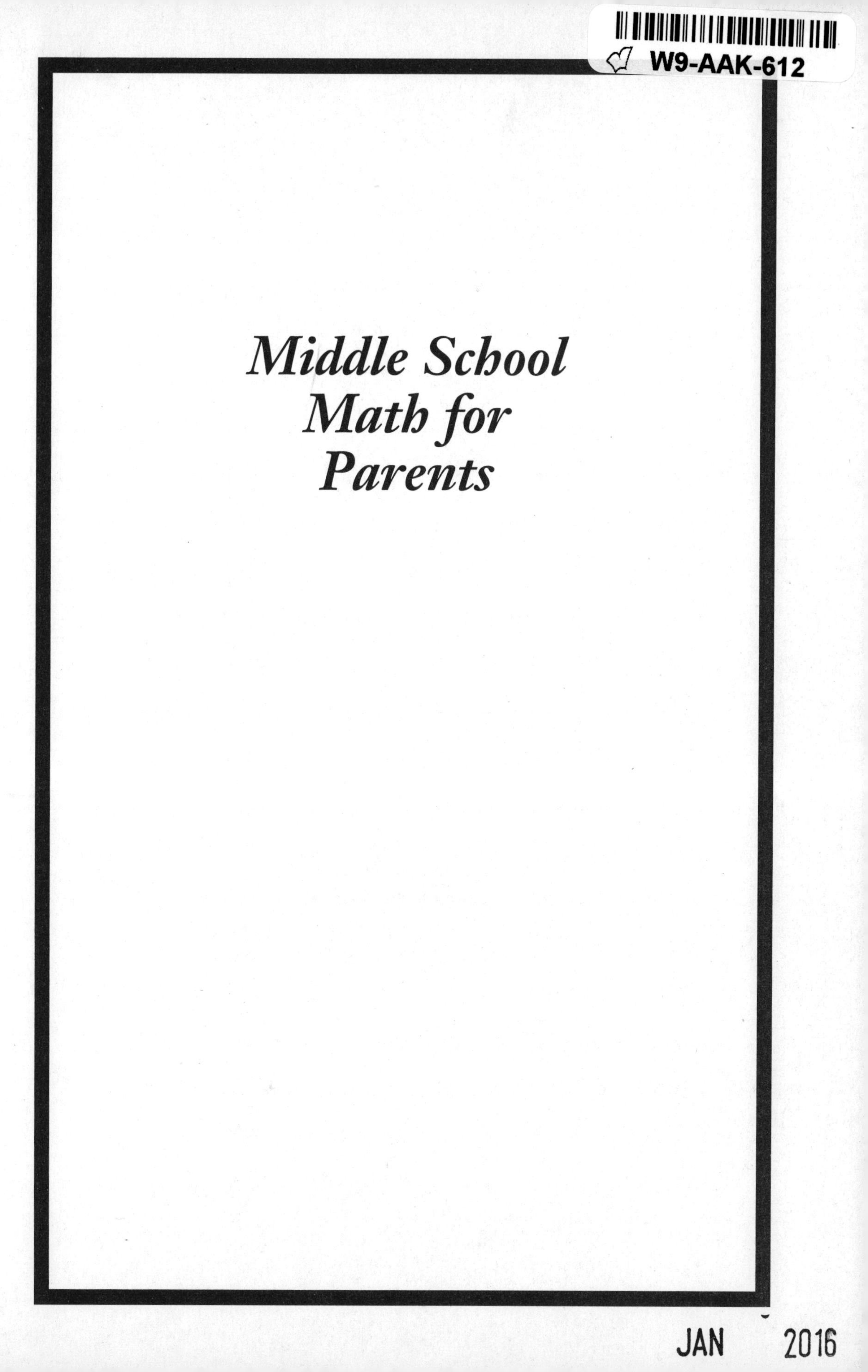
Middle School
Math for
Parents

Middle School Math for Parents

10 Steps to Helping Your Child Master Math

Scott Meltzer

Published in the United States by LearningExpress, LLC, New York.

Cataloging-in-Publication Data is on file with the Library of Congress.

Printed in the United States of America

9 8 7 6 5 4 3 2 1

ISBN 978-1-57685-944-5

For more information on LearningExpress, other LearningExpress products, or bulk sales, please write to us at:

80 Broad Street
4th Floor
New York, NY 10004

Or visit us at:

www.learningexpressllc.com

About the Author

Scott Meltzer is a National Board-certified early adolescence math teacher with 12 years of experience in education, both as a teacher and a coach (providing professional development and support for teachers). He has worked at all levels, elementary through college, in a variety of schools—public, alternative, and charter schools—in New York City, Seattle, and Philadelphia. His passion is working with inner-city students because he believes that all children can be great mathematicians and deserve a great education.

Meltzer lives in Philadelphia with his wife, Dr. Rachel Fleishman, and their two sons, Jacob and Benjamin. He dreams of writing the Great American Novel when he grows up.

Contents

Introduction

Raise your hand if your parent has ever tried to help you with math, and it ended in tears.

I have often posed this question to my eighth-grade students, with startling results. Even at that most self-conscious age, when everybody is afraid to be teased and emotion and self-doubt are taboo, a majority of students raise their hands.

I imagine a familiar drama playing out in many forms in many homes far too often. The roles may be different, but the core issue is the same.

The child is working on math homework at the kitchen table while Mom prepares supper.

"Mom, what's two-thirds divided by one-sixth?"

"You know I'm not a math person. You'll have to ask your father when he gets home."

When Dad walks in, the child approaches him with the homework.

"Dad, can you help me with this?"

"Sure, let's take a look."

But as Dad flips through the pages of the textbook, he finds himself disappointed and frustrated. The examples he was hoping for are not there—in fact, there don't seem to be very many numbers at all, for a math book. It's all words and pictures. *"Math isn't what it used to be. What was wrong with the old way?"*

Dad comes up with what seems to him like a reasonable solution: *"Let me show you the way I remember doing it."*

Dad works through some of the homework problems for himself. But the answers don't seem right, and when he types them into the calculator it keeps giving him these weird decimals. However, certain he has remembered the method correctly, he forges ahead and shows the child how to do it his way.

As the child tries to follow Dad's example, things don't work out. The child is doing it all wrong, and Dad is losing patience. Infuriated by the child's apparent *refusal* to understand, Dad's words and tone become less and less kind, until the child feels he is being outright aggressive. The child cannot take another second of Dad's impatience, harsh feedback, and inability to explain. Mom tries to stay out of it, but soon sparks fly. Once again, math has started a war.

If this scenario sounds familiar to you, you are not alone. The math war rages nightly in homes everywhere. Parents want to help and they try so hard, with the best of intentions. But it never seems to work. Some parents, like the mother in our little drama, believe that the best thing is to stay out of it. Others, like the father, want so badly to be helpful, but things always seem to go south.

The father in our example is aware of a generation gap that exists between today's students and their parents, with respect to the math classroom. Math education has changed in the last 10 to 20 years, leaving many parents more confused than ever.

It's a good thing you've found this book.

The Day My Mother Tried to Help with Calculus

There is no denying that some people learn math faster than others. That's not to say that anybody is incapable of learning high-level mathematics. There are just some people who get it quicker.

Growing up, I was always one of those people. New concepts clicked for me, and I was always able to repeat the procedures my teachers showed me.

But if you go far enough in any endeavor, you eventually hit a wall. For me it was the Fundamental Theorem of Calculus, in 12th grade.

$$\int_0^x f'(t)\,dt = f(x)$$

This is one of the most important equations in the history of mathematics, and one of the most beautiful. But for me in 12th grade, it was an indecipherable nightmare. I remember working on the living room floor, reading those textbook pages over and over, trying to make sense of it until I wanted to throw the book out the window. I could not figure out what that x was doing at the top of that integral because numbers were supposed to go there, not variables. It made no sense why there would be two different variables (the t and the x—I should count myself lucky that I knew at least what the f and the d meant).

Sensing my frustration, my mother came into the room and asked what was up. In language that in earlier years would have caused her to give my mouth the soap treatment, I expressed my frustration with the impossibility of the math. She shattered all that negativity and tension with three simple words:

"Can I help?"

I erupted instantly, not with rage, but with laughter. My mother? Helping me with math?

She doesn't remember the incident, but we both believe there is no way she could have been serious. About ten years later, I would go on to teach my mother the "lattice method," an algorithm for multiplying multi-digit numbers like 24 × 78, and she would tell me she never understood place value or multiplication before. Suffice it to say, my mother has never been a "math person."

Looking back at that incident with years of teaching experience, I now understand that there was a lot my mother could have done for me at the time. Even the most helpless of parents can be a huge help when it comes to math, and often the parent who barely knows how to add can be a better tutor than a parent with a Ph.D. in math. This book will show you why that is true and how you can help, no matter what your educational and mathematical background.

By the way, I understand and love the Fundamental Theorem of Calculus now—one of the greatest works the human mind has ever conceived.

Who This Book Is For

This book is for every parent or adult who wants to help a child with math.

In my years of teaching, I have spoken with countless parents who express the same frustration in myriad different ways: *I don't know how to help my child with math*. Your relationship with math is very important because it affects your approach to helping your child. There are so many different kinds of relationships that parents have with math. Some get it, some don't. Some like it, some don't. Some see its importance, while others don't feel we need it. Some parents wish they had teachers who used today's approaches, while others wish their children were in classrooms more like the ones they experienced.

This book is designed to help all parents learn to help their children with math, including:

- the parent whose heart speeds up when his child asks for help on math homework
- the parent who wants to understand why math these days is more about reading and writing than numbers
- the parent who can't understand why her child doesn't get it when it's so simple
- the parent who thinks it's okay if her child isn't a math person
- the parent who feels helpless, reliving her worst nightmares as her child experiences them
- the parent who knows the math well, but just can't explain it
- the parent who never understood this stuff in the first place
- the parent who used to help with math every night, but doesn't get this middle school stuff for some reason
- the parent who isn't sure how to ask the child's math teacher for help
- the parent whose child has already surpassed his own math ability but still wants to help

If you're considering using this book, then you've probably found that you fall into at least one of those categories. Have no fear! You *can* and *will* learn strategies for helping your child.

What This Book Is and What This Book Isn't

First and foremost, this is a how-to book. The purpose of this book is to teach you ways to help your child with math, no matter what your comfort level is with the subject. The major topics in middle school math will be presented, but the coverage is not meant to be anywhere near the rigor or depth that your child will experience in the classroom.

The bulk of these pages will be about math, but this is *not just* a math book. We are not trying to make you a math expert. Instead, the goal is for you to understand enough about the math to be able to start a conversation with your child. From there, you will be able to access your tool belt of strategies, which we will discuss and hone throughout the book. As you will see, knowing the math is not crucial to effective tutoring. I will show you the ways my mother could have helped me with my homework, despite not knowing the math, and the same methods will work for you.

The Common Core

You may have heard of the **Common Core State Standards**. The Common Core is a new set of math standards (instructions to schools and teachers about what they should be teaching) that has been adopted by most states. The math content that we will cover sticks to the Common Core for grades 6 through 8, with the occasional 5th-grade topic added to improve your understanding of current teaching methods. No attempt is made to follow the sequence that your child will experience in school or to identify when your child is likely to see certain topics, because even with the Common Core, every state, school, and classroom is a little bit different. For that reason, most chapters will cover topics from more than one grade level. For instance, this book has one data and statistics chapter, which includes topics from all three grades.

I would also encourage you to use the Internet as a source of more information, examples, lessons, and practice problems. There are countless great teachers who regularly post lessons on YouTube. Find a few that you like and refer to them often.

Chapters

Each chapter in this book will focus on a different tutoring strategy that you can use with your child. As we discuss each strategy, you will see how you can apply it while helping your child with a specific topic or concept in middle school math. The explanations of the content are designed to give you an entry point when helping your child with the math, and also to show you the way the math is currently presented in most middle schools.

Chapter 1: Dealing with Math's Positives and Negatives

This chapter deals with the most important idea in this book: A positive attitude toward math is critical to success. As we work through the concept of integers (positive and negative whole numbers), we will explore many ways that a negative attitude toward math can become an obstacle for your child, and we will develop strategies for creating and maintaining a positive attitude—for both you and your child.

Chapter 2: Fractionphobia: Facing the (Totally Rational) Fear of Rational Numbers

If attitude is the biggest factor of math success, fear is a close second. Many students are intimidated by math, a fear that for many starts right around the time they are learning about fractions. Not coincidentally, students who fear math almost always have at least one parent who fears math. This chapter will take a gentle approach to both fractions and the fears they inspire in even the most courageous among us.

Chapter 3: Mental Math Rules! Understanding Decimals by Unlocking the Secrets of the Base-Ten System

Many students believe they understand decimals better than fractions. In this chapter, we will develop your understanding of the base-ten system first with whole numbers, and then with decimals, so that you will be able to help your child develop confidence in mental operations.

Chapter 4: Learning It the "New" Way: Why Ratios Make More Sense Now Than Ever Before

This chapter is great for parents who understood math when they were in school, but don't get the new ways it's being taught. We will explore ratios and proportions, two of the most important concepts in all of middle school math, while giving you an understanding of the methods today's teachers use and the rationale for those methods.

Chapter 5: Visualize Everything: How Simple Geometry Problems Can Make Everything Else Easier

In this chapter, we will cover some of the basics of geometry—area, perimeter, and scale problems. In the process, we will see many examples of how these simple concepts can be applied to problems in other areas of mathematics, like operations, fractions, and algebra.

Chapter 6: The Joy of Math: Having Fun with Geometry

Math is fun. Sadly, many people do not believe that. You may be one of them, but fortunately, this chapter will help you begin to see the joy in mathematics. If you are having fun doing math with your child, then success isn't far behind. We will have some fun doing problems with circles, angles, transformations, and the Pythagorean theorem.

Chapter 7: The Two Most Important Words in Mathematics: How Why *and* Because *Can Be the Keys to Unlocking the Language of Algebra*

In many families, the parent helping the child with math is the wrong one. This chapter will show you why. As we study the basics of algebra—writing expressions and equations using variables—you will learn the power of the words *why* and *because*, two of the best teaching tools you will ever have.

Chapter 8: Solving Equations: The Many Meanings of "I Don't Know"

Good teachers know that "I don't know" can mean many different things. The challenge is interpreting it correctly as often as possible, and using it as a tool to help the student, your child. In this chapter, we will find the unknown in some classic "solve for *x*" problems as we discover effective ways to respond to your child's every "I don't know."

Chapter 9: Math Is Life: How to Find Algebra in Everything You Do, and Why It Matters

Remember $y = mx + b$? Linear relationships are a stumbling block for many students. Like fractions, it is a concept that can be the end of the line for some. In this chapter, we will apply many of the strategies we have already discussed as we study this difficult concept. We will see how prevalent linear relationships are throughout our daily lives, so that you and your child can both see the importance of math success.

Chapter 10: Life Is Math: How to Make Sense of Data and Statistics All Around You

Unlike linear relationships, data and statistics are math concepts you don't have to look too hard for. In the modern world, we are constantly bombarded with percentages, averages, and all kinds of statistics. In this chapter, we will learn the meanings of some of those things so that you and your child can interpret them with confidence, and separate the important ones from the rest.

Ready? It's Time to Do This!

By reading this far, you have already shown your commitment to becoming a great math helper for your child. Whatever your background, whatever your relationship with math, you have the power to make math a rewarding experience for yourself and your child. Stay calm, have fun, and don't let it get too stressful. You and your child will both do great. So let's get started!

1 Attitude Is Everything

Dealing with Math's Positives and Negatives

"Attitude is everything!" You've heard it said before, in many contexts. This message comes up so many times during our lives—maybe while looking for a new job or preparing for a championship game. The idea that we can improve ourselves with a tool as simple as attitude is pervasive in America—in fact, it's the basis of the American Dream. And yet, sadly, when it comes to math, many people lose sight of this simple and significant idea. So many people dismiss themselves with that alarmingly overused statement, "I'm not a math person."

If you are guilty of making that statement yourself, then you've come to the right place. You've come this far in life without your lack of math-confidence getting in the way. Or perhaps it has gotten in the way of the life you've wished for, but you've found detours around math obstacles. But now your child is struggling, and you fear he or she will have the same obstacles you've had. Or perhaps you understand that in the 21st century, your child will need more math than you've ever understood just to be able to make a living. You want to help, but how?

Why Your Child's Attitude Matters

By far, the most important factor in your child's future math success is how much time he or she spends practicing and studying. Contrary to what many people believe, some humans are not born more intelligent than others, or with a better "math brain." It's all about practice. The big question many parents face is how to motivate your child to practice as much as possible. There is a simple answer: It starts with your child's attitude.

Let's imagine two twin girls who are six years old—Alice and Betty. Their father takes them to play soccer for the first time. They play at the park for about an hour, just running around and kicking the ball toward the goal, which Dad is defending. Both girls have trouble controlling the ball, as one might expect. But by luck, on her first shot, Alice kicks it right in. Dad can't believe it, and for the rest of the day, Dad keeps telling Alice what a great shot it was.

Within a few weeks, Alice becomes a soccer nut. She wants to play all the time, and the practice makes her quite good. Betty, on the other hand, has no interest in soccer whatsoever, and finds Alice and Dad's common interest annoying, a family dynamic that will continue well into and beyond the girls' teenage years. Whenever Alice is at practice or a game, Betty finds herself with her sketchbook open in front of her, colored pencils at hand.

This simple example illustrates a cycle between attitude, practice, and performance. Good performance in an endeavor—be it sports, the arts, or math—helps one develop a positive attitude, which motivates the individual to practice more, leading to even better performance.

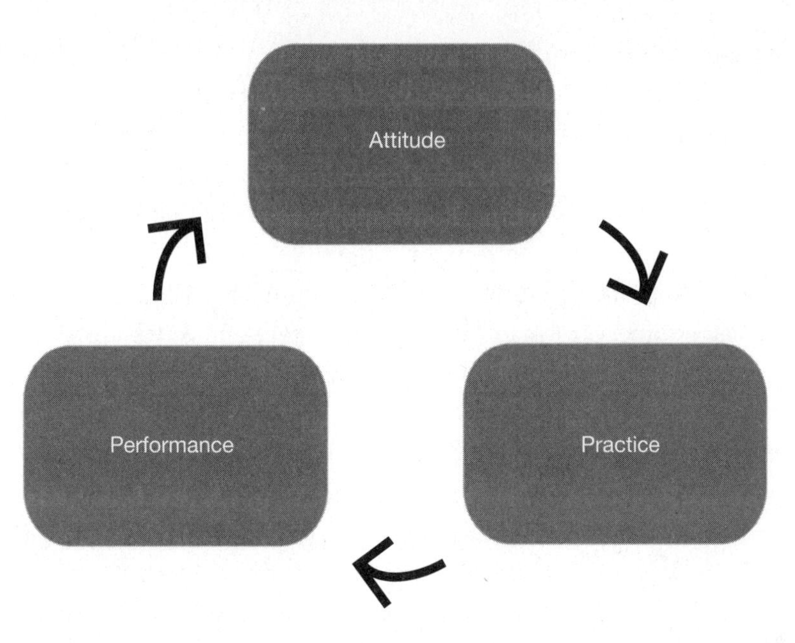

When we apply this cycle to math, attitude is the most important stage because it is where most breakdowns happen. Like the twins, if your child does not have a positive attitude toward math, she will not want to practice and will therefore not perform well, leading to more frustration and an even worse attitude. Things can spiral out of control.

THE CYCLE OF PRACTICE

We practice what we like, and we like what we are good at, so we practice it more, which makes us better at it, which makes us like it even more. Your child is going to have to do math, and being good at it is important. Therefore, a positive attitude is a must-have.

Why *Your* Attitude Matters

When was the last time your child said she wanted to be just like you when she grows up? If you have a teenager, it may not have happened for a few years. But if you think back, you can probably recall a time when you heard this pretty often. As parents, we are our children's first and greatest role models. They are born with a desire to emulate everything we do, a desire that remains strong for many years. Even as teenagers, they may say they hate everything about you, but they still cannot escape your influence. Sometimes that is a good thing. It makes teaching your children fairly simple—you can teach just about anything by *modeling*. If you want your child to say "please" and "thank you," then you say "please" and "thank you," and before long, your young child will learn to do the same.

But modeling can also cause trouble. My four-year-old son has recently begun using some very inappropriate language, and every time he does, my wife correctly scolds *me*, because she knows where he learned it. As fun and cute as it may be to teach little kids to use bad words, there is nothing amusing about a child saying he or she is not a "math person." If you do not consider yourself a "math person," then the best thing you can do to help your child is to avoid saying it. Because every time you say "I'm not a math person," you give your child permission to not be a math person, either. If instead you model a positive attitude toward math—an inquisitive, can-do attitude—then your child will learn to do the same. Think of it as an investment. It may be hard to invest a few positive words right now, but it will pay off later as the interest starts to grow.

Absolute Value: Turning a Negative into a Positive

Let's get into some math.

One tool that mathematicians often use is **absolute value**. Simply put, absolute value is a number's distance from zero on the number line.

7 is 7 units away from 0, so the absolute value of 7 is 7.

–7 is also 7 units away from 0, since distances are always positive (have you ever driven –7 miles?). So the absolute value of –7 is 7.

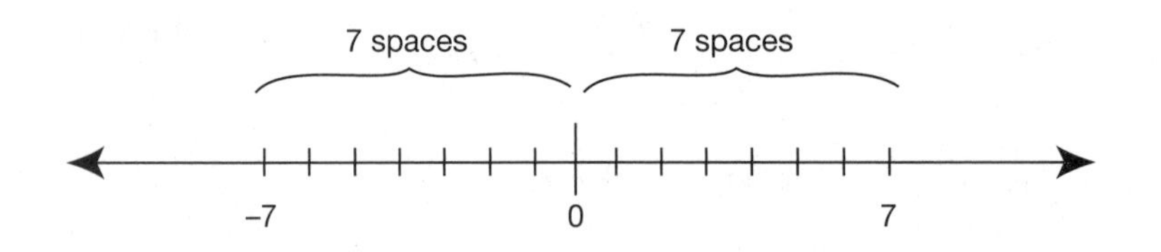

We use two bars to show absolute value:

- $|7| = 7$
- $|-7| = 7$

- $|148| = 148$
- $|-48.25| = 48.25$

Notice that absolute value takes every number and makes it positive, regardless of its original sign. If only we could do that to our attitudes! Unfortunately, we cannot, but we can learn to take a negative situation or statement about math and make it positive.

There are lots of ways to express difficulties in math while promoting a positive attitude. One thing to try is using "yet" statements. If your child says something negative about math, try using "yet" to make it positive. For example, "I'm not good at math" can be restated as "I don't understand this *yet*." The next time you or your child is feeling frustrated, practice using some of these.

Instead of . . .	Try saying . . .
"I'm not a math person, either."	• I know you can do this if you keep working at it. • This isn't easy, but you will get better at it. • Let's have some ice cream and then work on it some more together. • You weren't always good at baseball, either, but you kept practicing until you got it.
"Why can't you get this?"	• Let's see if we can figure it out together. • You know, it took the greatest mathematicians in the world centuries to figure this stuff out. So what if you don't get it in a day? • Let's see what mistakes you made so we can learn how to do it correctly. • I know you'll get it if you keep trying.
"I guess your brother is my math kid."	• I wonder if your brother could help you out a little. • If you practice some more, you'll get better at it. • There's nothing here you can't do—you just haven't gotten it yet.

Integer Addition: Take Control of Math Attitude

What do attitude, money, and numbers all have in common? Positives and negatives. Now that you understand how important having a positive attitude toward math is for both you and your child, let's use *integers*—positive and negative whole numbers—to find out how to make that positive attitude happen.

Attitude works just like a number line. On one end, you have the negative, and on the other end, the positive. At the center is 0, or a neutral attitude. It is neither positive nor negative. While a truly neutral attitude toward math is exceptionally rare, the number 0 is extremely common in math.

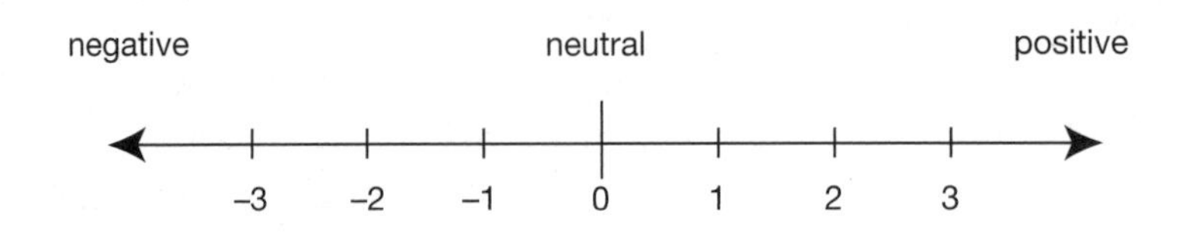

You may remember integers from your school days. Many people have trouble with integers because working with them has often been taught as a series of rules. This kind of teaching is usually ineffective because the rules usually don't make much sense and are therefore hard to remember. Students then make a lot of errors, which doesn't help anybody maintain a positive attitude toward math. Instead of listing and studying rules, we will practice using a few simple models for adding integers.

Integer addition is a great metaphor for attitude. Suppose your child has a negative attitude toward math. There are many things you can say to your child that may affect his attitude. For example, if your child brings home a poor test score, you might say, "What's wrong with you?" or you might say, "I know you'll do better next time."

The figure below shows what would happen if you choose the negative comment, "What's wrong with you?" Your child starts out with a negative attitude toward math, and the discouragement drives her even further into the negative, making a bad attitude worse.

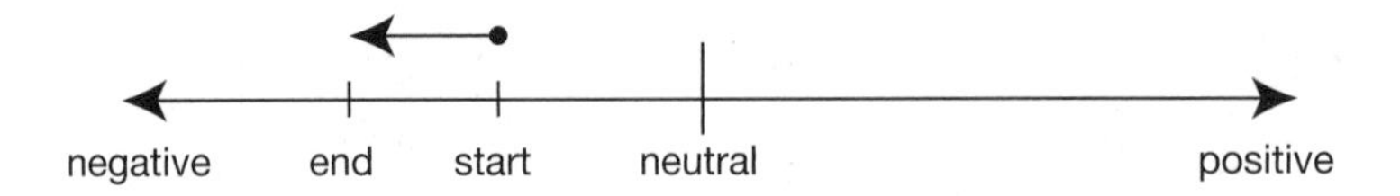

The next figure shows what would happen if you choose the positive comment, "I know you'll do better next time." Your child starts out with a negative attitude toward math, but the encouragement in your response makes him feel a little better, improving his attitude slightly. It might not be enough to make his attitude toward math positive overall, or even neutral, but *less negative* is a great first step. Turning a negative attitude into a positive one is a long process made of many baby steps, including the occasional step backwards.

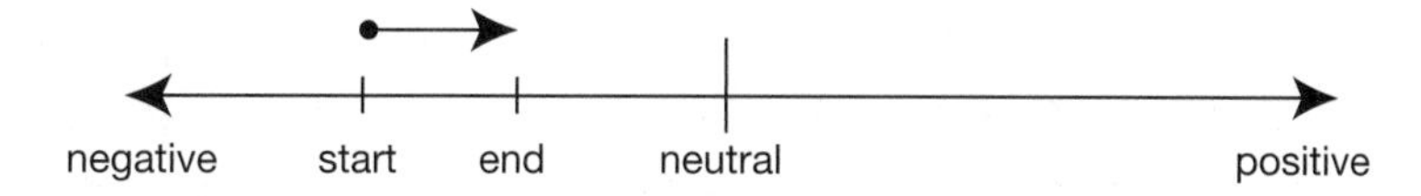

In both of the previous examples, the child's attitude had a start and an end. These are very useful tools for working with integer addition, using the basic equation

Start + Change = End

Let's look at a simple example with positive numbers: $7 + 2 = 9$. This means you start at 7 and add 2 (the change), and you end up at 9. This can be shown on a number line:

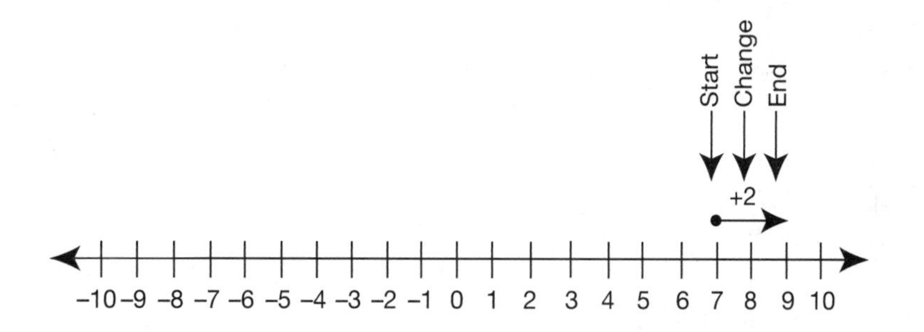

Adding integers is that simple: Find the start on the number line, draw the change arrow, and move to the end. When the change is positive, the arrow points to the right. When the change is negative, the arrow points to the left.

TERMS TO REMEMBER

- The **sign** of a number indicates whether the number is positive or negative. The sign of 8 is positive, but the sign of −12 is negative. 13 and 25 have the same sign, but −9 and 86 have different signs. 0 is the only number that has no sign.
- Many students find the terms **greater than** and **less than** to be confusing. **Greater than** (>) always refers to the number that is farther to the right on the number line (that is, *more positive*). So 9 is greater than 3. 16 is greater than −10. And −4 is greater than −11. This last example is the most confusing since −4 *may seem* smaller than −11. But remember that *greater than* refers to the more positive number. **Less than** (<) always refers to the number that is farther to the left on the number line (*more negative*).
- If a number **increases** it becomes more positive—in other words, you are adding a positive to it. A **decreases** means adding a negative, moving farther to the left on the number line, or becoming more negative.

Now let's look at money. An individual's finances are a constant stream of integer addition problems. We earn money, adding a positive. We spend money, adding a negative. If you have $10 but you spend $6, you have added –$6. If you owe somebody $4, your start value is –$4. If you then earn $7, your change is +$7, and you would end at $3. These examples are illustrated on the number lines below.

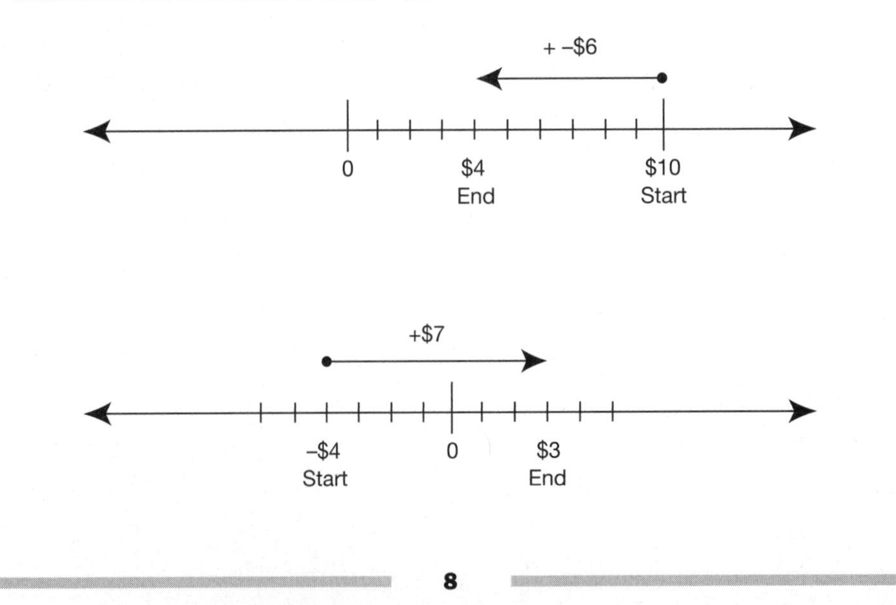

Examples

Here are a few more examples of how to use a number line to represent and solve integer addition problems. Try explaining to your child how to use a number line and the process "Start + Change = End," then ask your child to represent these problems or others on a number line for practice.

1. 4 + –3

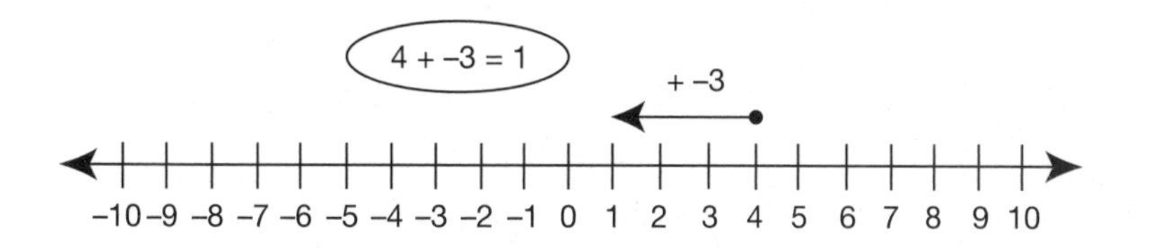

2. –8 + 6

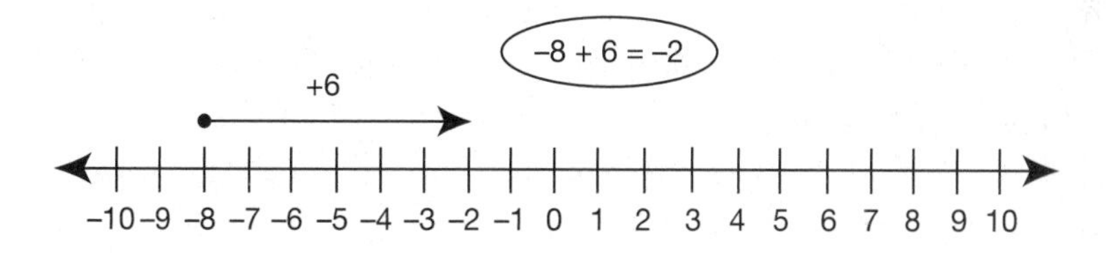

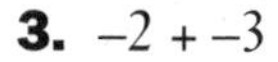

3. –2 + –3

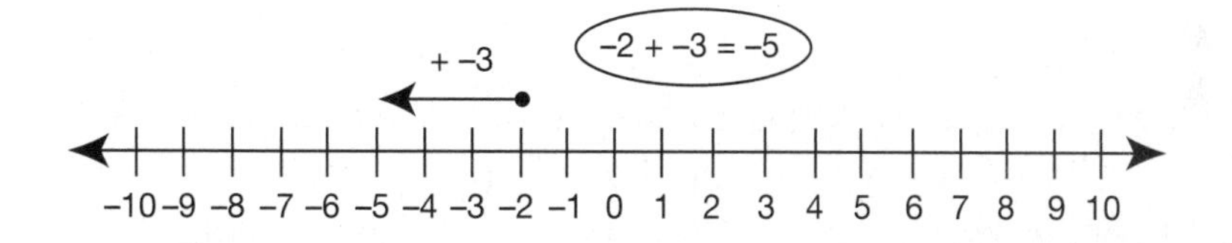

Notice that adding two negative numbers together gave us a larger negative number. If you owe one person $2, and you owe another person $3, you have a total of $5 of debt, or –$5. In other words, a negative statement will always make a negative attitude worse.

4. 3 + –8

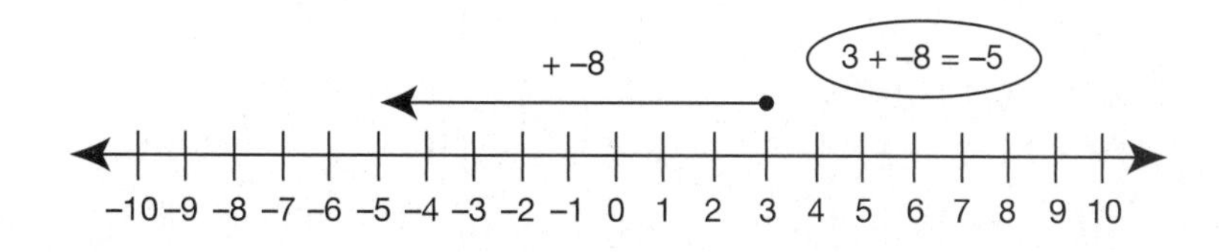

> The number line is a great visual tool for adding integers with small absolute values (that is, numbers close to zero). But what happens when you are faced with something like –436 + 192?

In a problem like this, it's not a great idea to draw the entire number line with marks all the way from 0 to –436. Instead, we can just use the number line to get us started thinking about the problem, which will often be enough to figure it out.

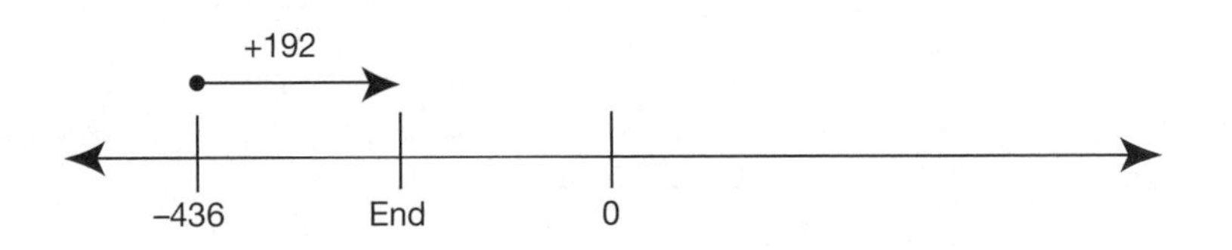

This number line is missing a lot of detail, but that's okay. We know that we need to start at –436, and that our change is in the positive direction since we are adding a positive number. Since 436 is larger than 192 (in terms of absolute value), the change will not be enough to bring us past 0 into the positives, so the end must be somewhere in the negatives between 0 and –436. To find out where the end is, simply subtract 436 – 192, which is 244. Why subtraction? That's a great conversation to have with your child—use the examples above to reason through it. If you think of it in terms of debt and cash on hand, it is very clear. Putting $192 toward a debt of $436 will still leave a person $244 in debt.

This illustrates why a negative attitude can take so long to change. As a parent, you will do and say many things to try to change your child's attitude about math (or anything else). If the starting attitude is a strong negative one, your efforts may be very small in comparison—after all, –3,000 + 1 = –2,999. Maybe it doesn't seem like the +1 did much to change that attitude. Be persistent, because it *was* a change, and repeating that change over and over will do a lot to get your child on the cycle of attitude, practice, and performance, so that he will begin to drive his own attitude in a positive direction.

Integer Subtraction: Don't Do It!

Life is tough, and there are many things out there that will try to ruin your day. It's easy to get caught up in all the bad things surrounding us. One way to stay positive is to actively avoid the things that will bring you down. Integer subtraction is one of them.

Subtracting two integers, like 5 – (–3) or –26 – (–8) can be very confusing for students.

Many teachers expect students to memorize rules for this process. To the teacher, this may seem simple, but in fact there is nothing simple about it. Depending on the teacher, these lists of rules can be quite long. Some teachers make students memorize songs to help remember the rules (just check YouTube). For most students, these teaching strategies *do not work.* Students may remember the song long enough to do okay on a classroom test, but by the next school year, they have forgotten it all. As an 8th-grade teacher, I have had arguments break out between my students year after year, where they are trying to agree on the rules that were taught last year. They say rules they remember, some right and some wrong. What concerns me the most is that the students who remember correctly are no more certain than the students who don't.

It's all very confusing, and the lack of confidence, combined with the constant errors that students will make, don't help improve anybody's math attitude—which is why it is best to just avoid subtraction completely. My 8th graders believe I'm allergic to it. Here's a story that shows how you can avoid subtraction completely.

An Integer Story

Anthony and Paul are best friends. They have everything in common, even their birthday. Their parents are close friends, too, so they often coordinate birthday parties and birthday presents. One year after their birthday, Anthony was disappointed with his birthday present from his dad. Feeling cheated, he asked Paul what he got from his father. "My dad gave me 25 bucks," Paul said.

"Lucky," Anthony sulked.

"Why, what did you get?"

"Well, I still owe my dad 60 bucks that he lent me for that video game. He said now I just owe him 35."

Is Anthony's disappointment justified? What happened, mathematically?

Paul was given $25. Whatever he started with, his change was simply +$25.

For Anthony, it's not so simple. We know his start was –$60, and his end was –$35. What happened in between? There are two ways to think about it.

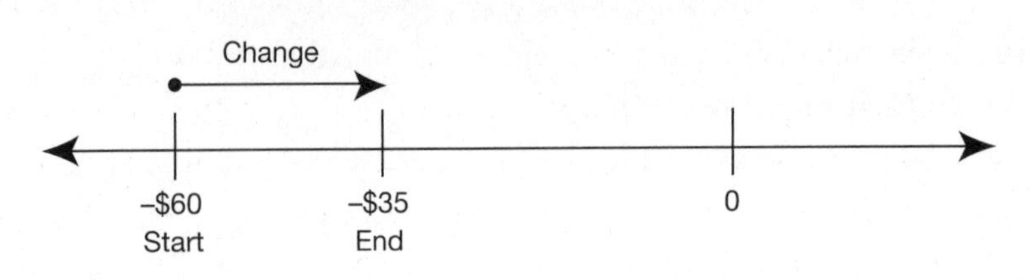

Using the number line, we can figure out that Anthony's change was +$25, the same as Paul's. But there's another way to think about it. In reality, Anthony's father didn't *give* him anything—in fact, his dad *took something away*. Specifically, he took away 25 dollars of debt. In other words, he subtracted a negative:

$$-\$60 - (-\$25) = -\$35$$

As the number line shows, this equation is mathematically equivalent to

$$-\$60 + \$25 = -\$35$$

So Paul and Anthony were both given the same amount of money.

The moral of the story? Subtracting a negative number is the same as adding its opposite.

The same thing is true about subtracting positive numbers: If you spend $19 in cash, you are subtracting a positive, but if you spend $19 on your credit card, you are adding a negative (a debt). Either way, you end up in the same place.

Now we have it—we can happily go through life without ever subtracting again if we don't want to. And our old friend the number line will always be there for us.

Let's work through a few examples of integer subtraction by converting everything (and I mean *everything*) to addition.

Examples

1. $17 - (-4)$
Subtracting a negative is like adding a positive, so we can change the expression to $17 + 4$, which is simple. $17 - (-4) = 21$. Subtracting a negative made the number larger.

2. –38 – 15

Subtracting a positive is the same as adding a negative, so –38 – 15 = –38 + –15. Our start is negative, and our change brings us further into the negatives, so our end will be even more negative:

–38 – 15 = –53

3. 12 – 5

This is a great example where students might think we're over-complicating things. That's because we are. Your child probably knew how to solve this problem in 1st or 2nd grade. When adding and subtracting more than two numbers at a time, lots of mistakes can happen if we're not careful, which is why we should practice this now.

12 – 5 = 12 + –5. So we are moving left on the number line, ending at 7:
12 – 5 = 7

4. –21 – (–9)

Yuck. This is the kind of problem that gives students the most trouble. But stick to the same ideas we've been using. First, change subtraction to addition. Losing a debt of $9 is like gaining $9, so –21 – (–9) = –21 + 9. Let's use the number line.

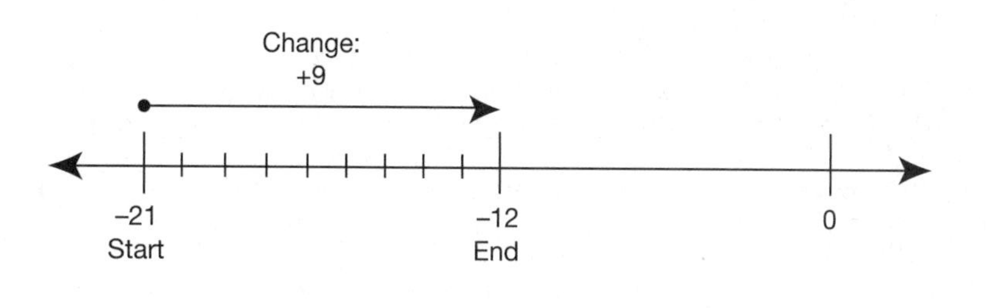

So –21 – (–9) = –12.

There are two things to notice about this number line.

First, there are no tick marks on the number line between –12 and 0. That's because we didn't need them. It's important for your child to learn to be efficient, and these number lines are a tool—they are the means to the solution, not the solution itself. You don't even really need the 0 on there, but it is helpful as a landmark.

Second, notice that the change arrow has +9 written on it. Whenever you or your child draws a change arrow, it is very important to note the operation. If the change arrow had –9 on it, instead of +9, it would be impossible to know whether it were a negative nine or subtracting nine. This distinction is very important to students.

Celebrating Success

Look back at how much we've done so far. Adding and subtracting integers is one of the most confusing concepts in middle school math. Whether you already had it down before starting this reading or you're just starting to understand it for the first time, you've made some progress. Even if you still don't feel like you get it, the effort is more important than the endpoint, and you've already come this far. Give yourself a pat on the back, take a coffee break, or have a candy bar—celebrate your success in a meaningful way.

The same goes for your child. Reward every effort, because you want to train your child to become better and better at practicing—practicing both the math and the positive attitude. If you hear your child say something positive about math, tell her, "It's great to hear you say that." If you see her putting a little more effort than usual into her math homework, point it out to her—"I see you're really working hard at that. Great job!" Small praise for small success can go a long way toward building a positive attitude.

Be a cheerleader for your child. Have you ever been to a football game where the cheerleaders told their losing team, "You suck, just quit"? Cheerleaders are great models for the kind of attitude-building work you can do as a parent. Improved math grade? "Great job, keep it up!" Still struggling? "Work hard, you'll get there soon!"

Multiplication with Signed Numbers

When your child first learned multiplication back in 2nd or 3rd grade, it was likely presented as **repeated addition**. Multiplication in all forms becomes simple if you can remember what it means: repeated addition. So what is repeated addition?

Examples

- Take 3×5 as an example. 3×5 means we are adding 3 groups of 5, say three $5 bills, or the fingers on three hands. In other words,

  $$3 \times 5 = 5 + 5 + 5$$

  This is what we mean by repeated addition: the 5 is being added repeatedly, in this case, three times.

- $7 \times 6 = 6 + 6 + 6 + 6 + 6 + 6 + 6 = 42$

- $6 \times 7 = 7 + 7 + 7 + 7 + 7 + 7 = 42$

You may remember the word **commutative**. The last two examples demonstrate the **Commutative Property of Multiplication**—fancy words for a simple idea, that 7×6 is the same as 6×7. It doesn't matter if you combine 7 groups of 6 items, or 6 groups of 7 items—either way, you have 42 items. This idea will be very important when we are multiplying integers.

What happens if we have to multiply 4×-6?

Because of the Commutative Property, there are two different ways to think about this problem. Either we are adding 4 groups of –6, or we are adding –6 groups of 4. Hopefully one of these repeated additions makes more sense to you. How would we draw –6 groups of 4? It's difficult to visualize. But we can easily add 4 groups of –6:

$$4 \times -6 = -6 + -6 + -6 + -6 = -24$$

If I owe $6 to four different people, I am $24 in debt altogether.

MAKING SENSE OF PROCEDURES

All those rules you have probably forgotten? "A negative times a positive makes . . ." and all that? No need for it. Simply remembering that multiplication is repeated addition makes it all simple. Math is built on a foundation of definitions, not rules. If your child can remember what multiplication means, and really apply that definition to any example, then she will never need to rely on an esoteric rule.

So we know how to multiply a positive times a positive, and a positive times a negative (or a negative times a positive). But what about a negative times a negative?

Multiplication tables can help us make sense of a problem like -2×-3. Let's look at the multiples of -2:

-2×1	-2
-2×2	-4
-2×3	-6
-2×4	-8
-2×5	-10
-2×6	-12

In each row, we are adding -2 to what came before (remember, this is repeated addition), so the product decreases by 2. But we can also think of it the opposite way: Each time we go *up* a row, multiplying by one *less*, the product *increases* by 2. Mathematicians love patterns that can extend forever in either direction. So let's extend the top of the table with the factors 0, -1, -2, and -3:

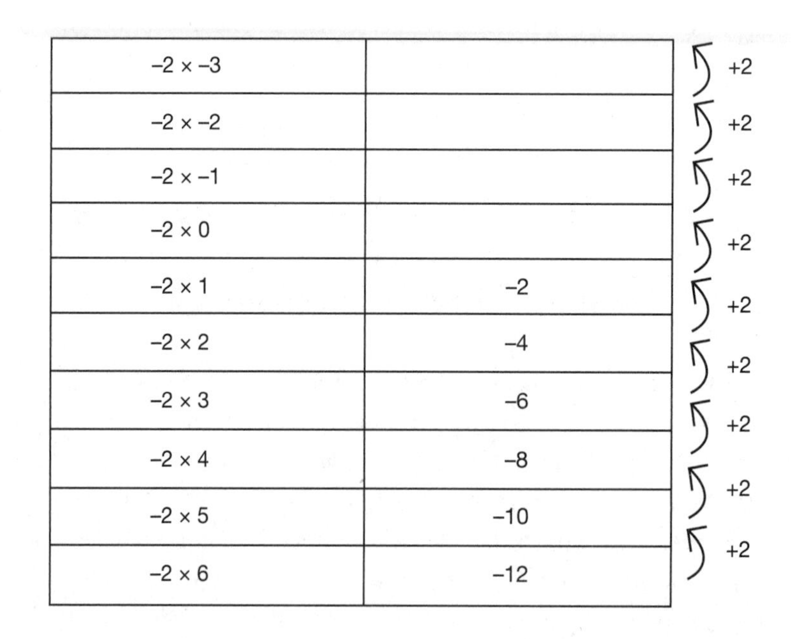

-2×-3	
-2×-2	
-2×-1	
-2×0	
-2×1	-2
-2×2	-4
-2×3	-6
-2×4	-8
-2×5	-10
-2×6	-12

By continuing the pattern, we can complete the table:

−2 × −3	6
−2 × −2	4
−2 × −1	2
−2 × 0	0
−2 × 1	−2
−2 × 2	−4
−2 × 3	−6
−2 × 4	−8
−2 × 5	−10
−2 × 6	−12

So we can see that -2×-3 is *positive* 6. When a negative number is multiplied by a positive, the product is negative, but when a negative number is multiplied by a negative, the product is positive. Rather than depending on this idea as a rule somebody told us, we have justified it using a very simple number pattern and the definition of multiplication.

Examples

1. $-5 \times 8 = -40$

2. $13 \times -3 = -39$

3. $-6 \times -8 = 48$

Using Fact Families for Multiplication and Division

When we discussed subtraction, we found a simple way to turn every subtraction problem into an addition problem. The same can be done for division, which is even more difficult for students than subtraction. Addition and subtraction are **inverse operations**, which means they do opposite things to numbers. That's why we can convert subtraction to addition by changing the sign of one of the numbers. Similarly, multiplication and division are inverse operations, so we can convert a division problem into multiplication if we are careful about how we do it. We will use another model, called **fact families**.

Your child may be familiar with fact families from elementary school. The idea is to take three numbers that are related by multiplication and division (or addition and subtraction—it works the same way), and to write four different equations with those three numbers. For example, here is the fact family relating 3, 4, and 12:

$4 \times 3 = 12$
$3 \times 4 = 12$
$12 \div 4 = 3$
$12 \div 3 = 4$

Notice that the first two equations are simply an expression of the Commutative Property, which is the idea that in multiplication (or addition), the factors can be written in either order, and the product does not change. The third and fourth equations show that since 12 is the product of 3 and 4, it can be divided by either factor to get the other factor.

You may find it helpful to write a few more fact families with simple whole numbers. Choose any multiplication equation and write the other three equations in the fact family, then check that they are all true equations.

Consider the numbers 8, –5, and –40, which make up a multiplication/division fact family:

$8 \times -5 = -40$
$-5 \times 8 = -40$
$-40 \div 8 = -5$
$-40 \div -5 = 8$

If you find this fact family confusing, relate it back to a simple numerical example like $3 \times 4 = 12$.

Now that we understand fact families, we can use them to solve many problems, beginning with integer division problems.

TERMS TO REMEMBER: DIVISION VOCABULARY

There are three important but often forgotten words in division: *divisor*, *dividend*, and *quotient*. In the equation:

284 ÷ 4 = 71

284 is the **dividend**.

4 is the **divisor**.

71 is the **quotient**.

In long division form, the *dividend* is the number that goes inside the box. In fraction form, the numerator is the dividend, and the denominator is the divisor.

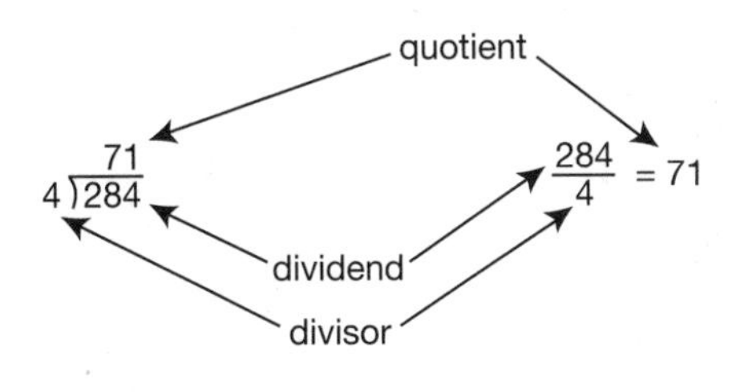

Examples

1. –30 ÷ –6

Let's use a fact family, putting a ? in place of the unknown quotient:

–30 ÷ –6 = ?
–30 ÷ ? = –6
–6 × ? = –30
? × –6 = –30

We can use either of the last two equations and the idea of repeated addition to find the value of the ?. How many times would we add –6 to make –30? Since 6 × 5 = 30, it takes 5 groups of –6 to make –30. Therefore –30 ÷ –6 = 5.

2. $36 \div -9$

Like in the previous example, we can make the entire fact family:

$36 \div -9 = ?$
$36 \div ? = -9$
$-9 \times ? = 36$
$? \times -9 = 36$

Adding –9 repeatedly won't give us a positive number, but as we saw earlier, we can move up the multiplication table, using negative factors. We know that $4 \times 9 = 36$. So $-4 \times -9 = 36$. Therefore, $36 \div -9 = -4$.

3. $-100 \div 25$

Now that we've practiced a few times, let's do this one without the whole fact family, since we really only need one of the multiplication equations.

$-100 \div 25 = ?$
$25 \times ? = -100$

Since $25 \times 4 = 100$, we know that $25 \times -4 = -100$. Therefore, $-100 \div 25 = -4$.

Summary

In this chapter, we have discussed strategies for cultivating a positive attitude toward math for both you and your child. Remember the most important points:

- It is not O.K. to be O.K. with not being a "math person." You will never encourage your child to work hard at math unless you show that you believe success can be achieved for both of you. Your child will be what you model.
- Practice!
- Maintain your absolute values: Use positive words whether your child is doing well or not. If you tell him he'll do better next time, or you tell him he's never going to get it, most likely you will be right.

- The more your child experiences success, the more she will want to work hard for further success. Celebrate all victories, big and small, so that your child can learn to see the successes for herself. If your child had a bad test score, she may be beating herself up as a failure. Find the problems she did well on and focus on that. Be a cheerleader, and remind her that she worked hard to get ready for that test and she'll do better next time.

We have worked extensively with integers. Hopefully you and your child both understand integer operations better now than you did before (and if so, a high-five or a hug is in order!). Remember that the rules don't always help—instead, stick to the tools:

- Use the number line for addition.
- Use addition for subtraction.
- Use repeated addition and extended times tables for multiplication.
- Use fact families to turn division into multiplication.
- Use money to contextualize any problems you can't understand—it will usually help you make sense of the problem.

In Chapter 2, we will look at another aspect of math attitude, which is mathphobia, the fear of math. As we explore strategies for dealing with mathphobia, we will extend our understanding of the number system to include fractions, the most frightening numbers of all to many students.

Fractionphobia

Facing the (Totally Rational) Fear of Rational Numbers

The First Day of Middle School

Tyshawn is a new 6th-grade student. He can't wait to start middle school—the freedom, the new friends to make, all those different teachers. The first day starts out great. Waiting in the commons that morning, he gets to hang out with some buddies from elementary school who he hasn't seen all summer. They notice how everyone is taller. They wonder what classes they will have together. Then it's time for homeroom. The teacher is cool—strict, but fun—and she shows them how to open their locks and how to read their schedules. Then Tyshawn's heart stops. He has math first period.

When the time comes, he walks to room 213, remembering what his mother said that morning: *Make me proud.* For the briefest moment, he cannot bring his foot to cross the threshold into that room. Tyshawn knows nothing about the teacher in that classroom or who his classmates will be. He is terrified of math. For that entire period, he sits on his hands, telling himself he is cold as a way to justify the chills and shivers he cannot stop. He isn't listening to a word the teacher is saying. Instead, he is rehearsing different wisecracks in his head, so that if and when she calls on him, he will be ready. He would rather get in trouble than embarrass himself with a wrong answer.

Far too many middle- and high-school students feel this way every day in the math classroom. Tyshawn is no different from countless peers in all kinds of schools and communities across the country. But for most of those students, it wasn't always that way. Tyshawn used to love math—and he used to be quite good at it. He still keeps certificates and ribbons from 2nd- and 3rd-grade math competitions up on the wall in his bedroom.

So what changed for him? What could possibly have turned his favorite subject into his worst fear?

The Origin of Mathphobia

Psychologists define a phobia as an irrational fear of something specific. You've probably heard of claustrophobia, the fear of tight spaces, or arachnophobia, the fear of spiders. Glossophobia, the fear of public speaking, is very common. Phobias, while considered irrational, usually have an origin somewhere in a person's early life. Mathphobia is very common among students in the middle grades and older, and its origin is usually in a more specific fear: the fear of fractions.

It starts around 4th or 5th grade. As a parent, you are well aware that up to that point, math is simple—adding, subtracting, multiplying, dividing, all with just plain, old "normal" numbers. Nothing fancy, no fractions, no decimals, no negative numbers. Simple. Then in the last few years of elementary school, teachers start throwing fractions out there, and everything goes haywire.

Take a fraction like $\frac{3}{5}$. What the heck does that mean? What do those two numbers have to do with each other? And you want me to *add* that with another fraction?

Fractions are hard—certainly harder than anything your child encountered before. And there are lots of other struggles that come with them. For one thing, many elementary teachers have trouble with fractions themselves, and therefore don't teach the concepts as well or as thoroughly as other subjects. Perhaps your child's teacher didn't have much patience for your child's struggles with fractions and said an unkind word . . . the kind of word that sticks with kids and grows into a full-fledged fear. Perhaps your own confidence in fractions is not so great, so maybe you weren't able to help out as much. This is certainly not your fault, and you are doing the right thing by looking for answers now.

FRACTIONS: THE GATEWAY TO MATHPHOBIA

Whatever the cause, mathphobia is very real, and very natural. For most students who struggle with middle school math, that difficulty can be traced back to fractions, the first time many students get the idea *I'm not good at math.* If your child's fear of fractions can be cured, his fear of math will likely vanish as well.

There Is a Cure

Fortunately, most phobias are curable. It takes time, and there may be lots of backtracking along the way, but there is a solution. Experts say that if you have a phobia that is affecting your life in a negative way, you must face that fear in controlled circumstances in order to eventually overcome the fear.

Take dogs, for instance. When your child was little, did he ever get frightened when encountering a dog, large or small? How did you approach this situation?

There are three approaches that I see parents taking when I am out walking my beagle:

1. Run and hide! Get the kid away from the dog and the dog away from the kid as soon as possible.
2. Gentle introduction. The dog owner holds the dog securely, and the parent greets the dog, showing the child that the dog is harmless and encouraging the child to try petting the dog gently.
3. Trial by fire. The parent leaves the child to his own devices, while the dog is free to approach the child in any way it sees fit.

Option 1 may seem like the most logical. If you're scared of something, stay away from it. This may work just fine when you're talking about dogs—they are easy enough to avoid—but when it's something you need in life, like math, option 1 is not the best approach.

Option 3 is not good for anybody. In the case of the dog, it could even become a dangerous situation, as the child's reaction will be unpredictable and may drive the dog to behave irrationally, particularly if it is a dog you do not know well.

Option 2 is the only way to help the child overcome the fear. By showing the child that the dog will not hurt him, you make the child more likely to approach the next dog he sees, and the next, and the next after that. As we mentioned when discussing attitude in Chapter 1, small changes make a big difference over time.

A Gentle Approach to Fractions

Now that we understand how to cure the phobia, let's take it head on with a gentle approach to fractions.

The approach we will take follows the Common Core standards, which are now in place in many states. In the Common Core, fractions are formally introduced as early as 3rd grade, with concepts building slowly on each other year after year. While all students may not have had the benefit of this approach, it should help you as a parent to reduce fractionphobia for yourself and your child.

The most important thing to understand about fractions is that they are numbers. They may not look like regular numbers, and they don't always behave as nicely as whole numbers, but fractions are numbers. They can do all the things that numbers do and can be used in all the same ways. While fractions appear to be made up of two numbers, a fraction is just a single number. Later in this chapter, we will discuss the relationship of those two numbers.

REMEMBER

A fraction is a number.

Unit Fractions

We begin with the most basic kind of fractions, called **unit fractions**. Unit fractions are fractions that have a 1 on the top, which is known as the **numerator**. Examples of unit fractions are 1/2, 1/3, 1/4, etc. These may also be written in the following form: $\frac{1}{2}$, $\frac{1}{3}$, $\frac{1}{4}$, etc. These are read as one-half, one-third, one-fourth, and so on. $\frac{1}{29}$ is read as one-twenty-ninth.

When you see a unit fraction, it means that something whole has been cut into equal parts, and the unit fraction represents *one* of those equal parts. The **denominator**, or bottom number, tells you how many parts there are. So if your child eats $\frac{1}{4}$ of a candy bar, that is one of four equal-sized pieces.

Examples

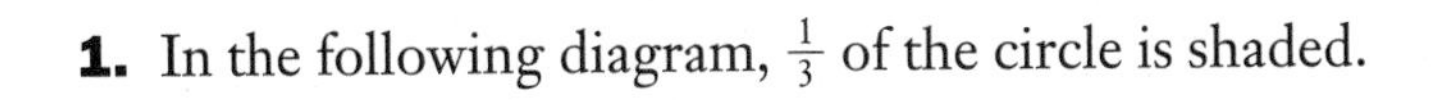

1. In the following diagram, $\frac{1}{3}$ of the circle is shaded.

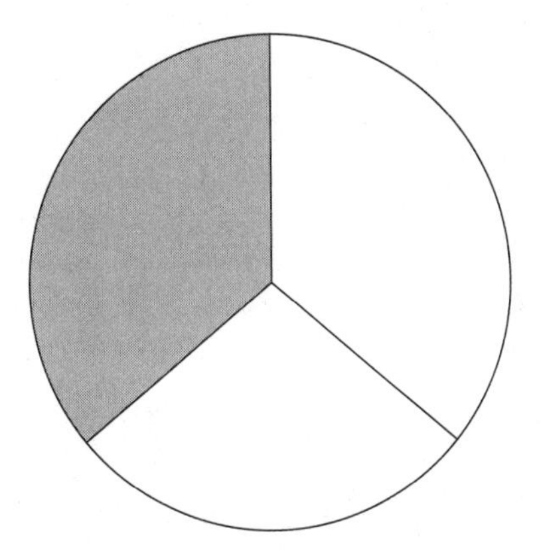

2. $\frac{1}{5}$ of the rectangle below is dotted.

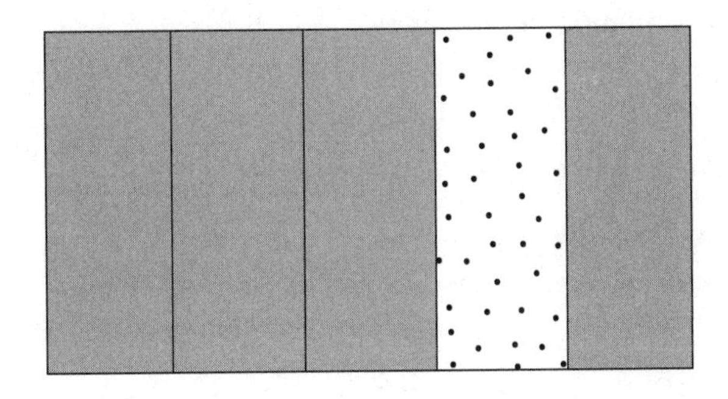

3. If you cut a pizza into 8 equal slices, each slice is $\frac{1}{8}$ of the whole.

4. 25¢ is $\frac{1}{4}$ of a dollar.

5. **One non-example:** In the following diagram, the shaded region is *not* $\frac{1}{4}$ of the shape. Why not? The pieces are not equal in size. If a family of four were sharing a triangular cake, would anybody be satisfied with that piece?

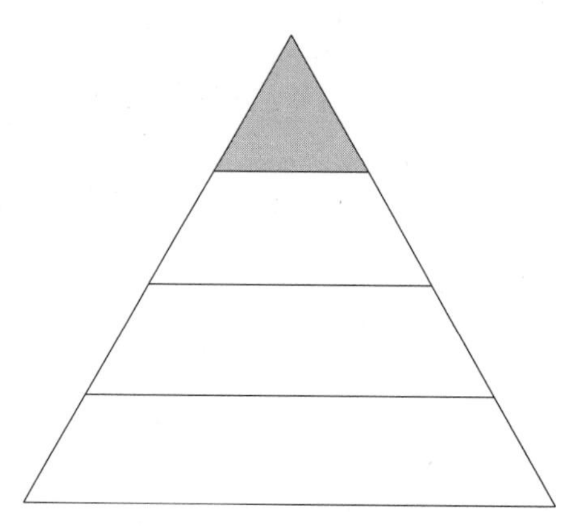

Non-Unit Fractions and Mixed Numbers

Of course, not all fractions are unit fractions. Sometimes we eat more than one slice of pizza. That's why fractions have a numerator *and* a denominator.

TERMS TO REMEMBER

The bottom number in a fraction is called the **denominator**. The denominator tells you how many equal pieces the whole has been cut into. The larger the denominator, the smaller the pieces, since the whole has been cut into more pieces.

The top number in a fraction is called the **numerator**. The numerator tells you how many of the pieces you are considering.

Examples

1. Suppose Elias has eaten $\frac{3}{5}$ of his candy bar.

$\frac{3}{5}$ ← Numerator / ← Denominator

The denominator tells us that the candy bar was cut into 5 equal pieces. The numerator tells us that Elias has eaten 3 of those pieces. So we understand that he has not eaten the whole bar yet, but he has eaten more than half of it.

2. In the following diagram, $\frac{2}{4}$ of the circle is shaded.

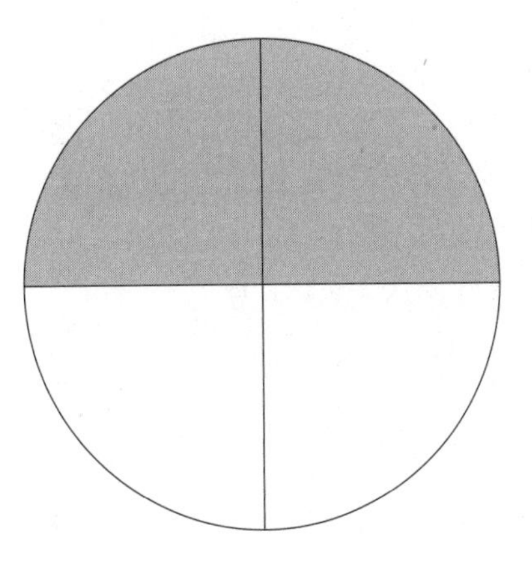

3. $\frac{5}{8}$ of the rectangle below is shaded.

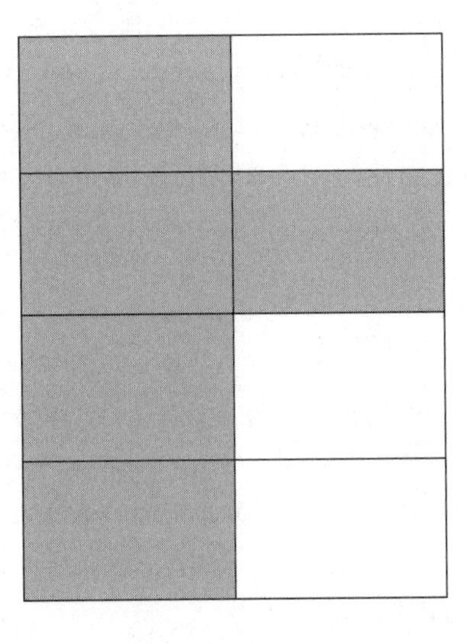

4. If you eat 3 slices of a pizza that is cut into 8 equal slices, you have eaten $\frac{3}{8}$ of the pizza.

5. 75¢ is $\frac{3}{4}$ of a dollar.

Fractions can also be used to indicate quantities larger than a whole. If you've ever bought pizza for a bunch of 8th graders, you know this well. Last time I had a pizza party for one of my classes, Nathaniel ate 9 slices. Each pie was cut into 8 slices, so Nathaniel ate $\frac{9}{8}$ of a pie. Even though the slices came from several different pies, 8 of those slices could fit together to make one pie. We can show the slices Nathaniel ate in the following way:

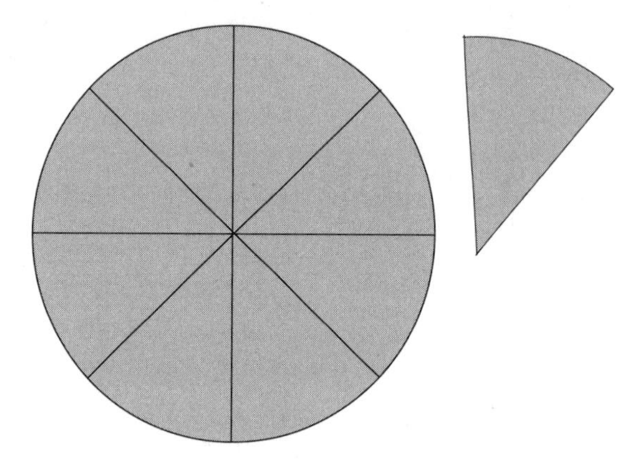

There are 9 slices shaded. We still have a denominator of 8, because the denominator tells us how many pieces make up a *whole*—in other words, the 8 tells us the size of the slices, with each slice being $\frac{1}{8}$ of a pie. Nathaniel ate 9 slices. When the numerator is larger than the denominator, we call that an **improper fraction**—not that there's anything wrong with it.

Another way to think of this is that Nathaniel ate 1 whole pizza and 1 extra slice. We can write this as $1\frac{1}{8}$, which we read as "one and one-eighth." This form is called a **mixed number** because it combines a whole number with a fraction. Nathaniel ate one and one-eighth pizza pies.

$$\frac{9}{8} = 1\frac{1}{8}$$

Examples of Mixed Numbers and Improper Fractions

1. It took Shanaia $3\frac{1}{2}$ days to drive from Los Angeles to Washington, DC.

2. $1\frac{2}{3}$ of the circles below are shaded. This is the same as $\frac{5}{3}$, since 5 pieces are shaded and each piece is $\frac{1}{3}$ of a whole circle.

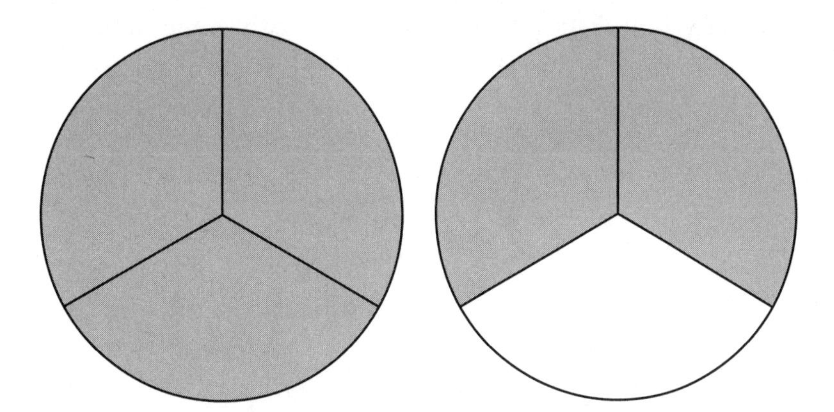

3. $\frac{7}{2}$ rectangles are dotted.

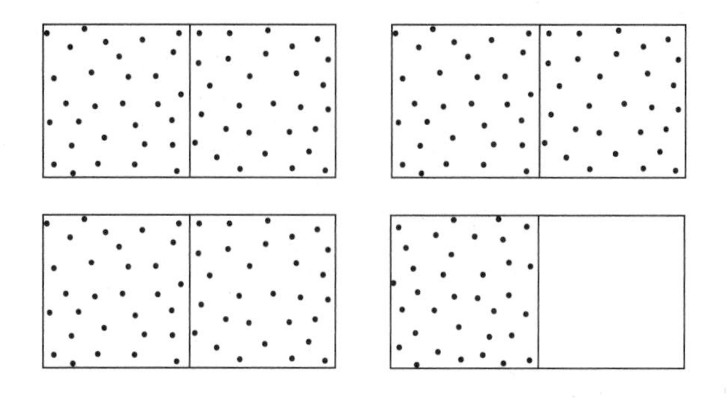

7 pieces are shaded in total, and each piece is exactly $\frac{1}{2}$ of a rectangle. To express this as a mixed number, there are 3 whole rectangles and 1 additional half shaded, so $3\frac{1}{2}$ rectangles are shaded. This is the same mixed number we saw in Example 1, so we could also say that Shanaia drove for $\frac{7}{2}$ days.

So far, so good. If you consider yourself fractionphobic, you've already come a long way, and hopefully the dog hasn't bitten you yet. The fact that the book is still open in front of you is a great sign.

Now we're ready to start operating on fractions. As always, we will make sense of the math as we go along, in small, gentle steps.

Equivalent Fractions

Sometimes different-looking fractions actually mean the same thing. These are called **equivalent fractions** because they have the same value.

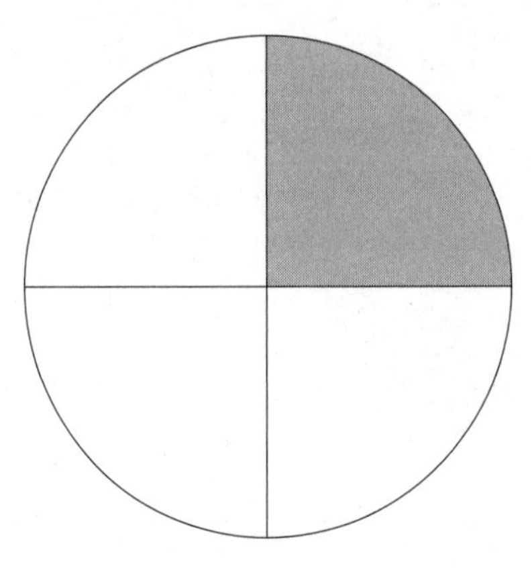

This pizza has been cut into 4 equal pieces. Each piece is $\frac{1}{4}$ of the pizza.

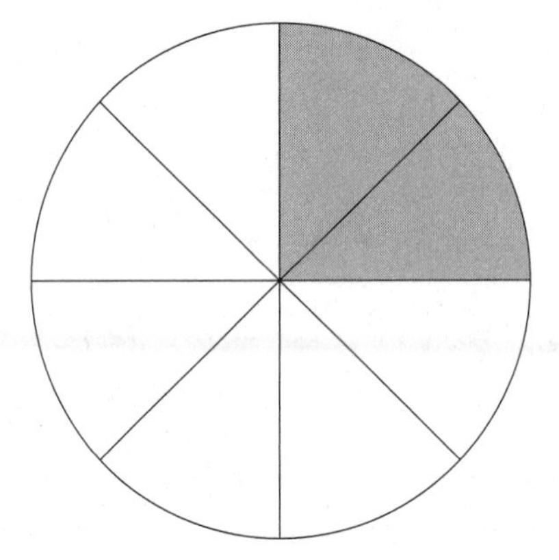

This pizza has been cut into 8 equal pieces. Each piece is $\frac{1}{8}$ of the pizza. In both of these pictures, the same amount of pizza is shaded. One slice out of 4 is equivalent to two slices out of 8. Therefore:

$$\frac{1}{4} = \frac{2}{8}$$

Notice that both the numerator and denominator have doubled from $\frac{1}{4}$ to $\frac{2}{8}$. **If we cut the pizza into twice as many pieces, you have to eat twice as many pieces to consume the same amount.**

It is important to know when a fraction is in **lowest terms**, or **simplest terms**. Lowest terms mean that any other equivalent fraction will have larger numbers for its numerator and denominator. $\frac{1}{4}$, $\frac{2}{8}$, $\frac{3}{12}$, $\frac{4}{16}$, etc. are all equivalent, but of all the fractions on this list, only $\frac{1}{4}$ is in lowest terms. Most of the time, your child will be expected to give answers that are in lowest terms.

Equivalent fractions are often easiest to understand with rectangular models.

Examples

1. The image below shows that $\frac{3}{6} = \frac{1}{2}$:

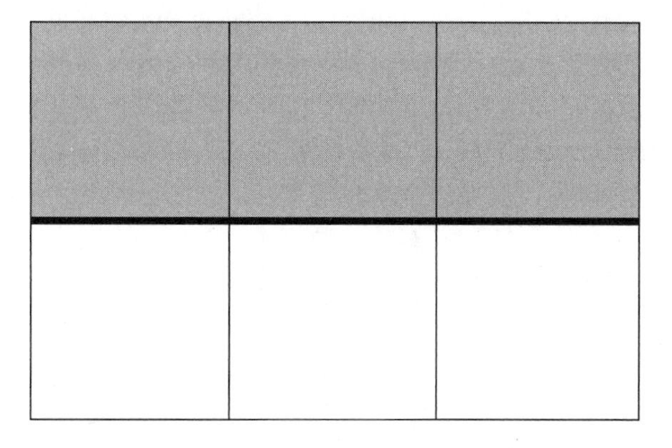

3 pieces out of 6 are shaded. 1 *row* out of 2 is shaded. Notice that $1 \times 3 = 3$ and $2 \times 3 = 6$.

$$\frac{1 \times 3}{2 \times 3} = \frac{3}{6}$$

If you cut the whole into 3 times as many pieces (6 pieces instead of 2), then you need 3 times as many pieces to make the same amount (3 instead of 1).

2. $\frac{9}{12} = \frac{3}{4}$

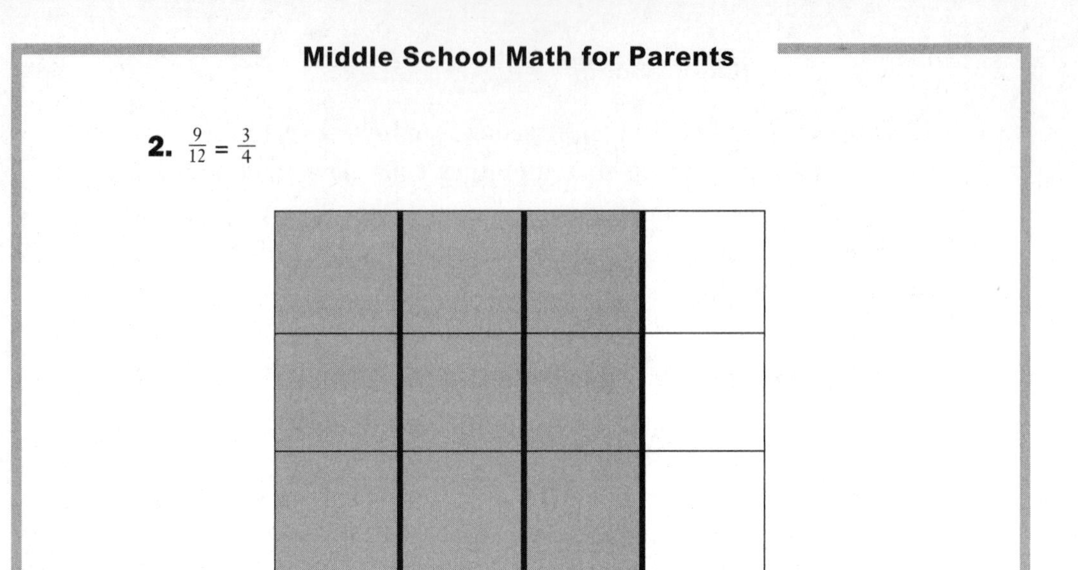

9 pieces out of 12 are shaded. 3 *columns* out of 4 are shaded. Therefore, $\frac{3}{4} = \frac{9}{12}$. Notice that $3 \times 3 = 9$ and $4 \times 3 = 12$.

3. Sometimes rows and columns don't work so well:

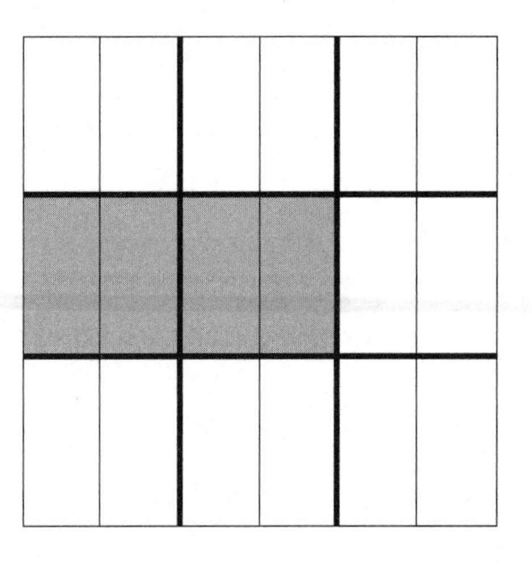

Here the bolder lines show the square cut into 9 equal pieces. The thinner lines have cut each ninth in half, creating a total of 18 pieces. $\frac{4}{18}$ are shaded. $\frac{2}{9}$ are shaded. Therefore, $\frac{4}{18} = \frac{2}{9}$.

4. What if you shade *all* the parts in a fraction picture?

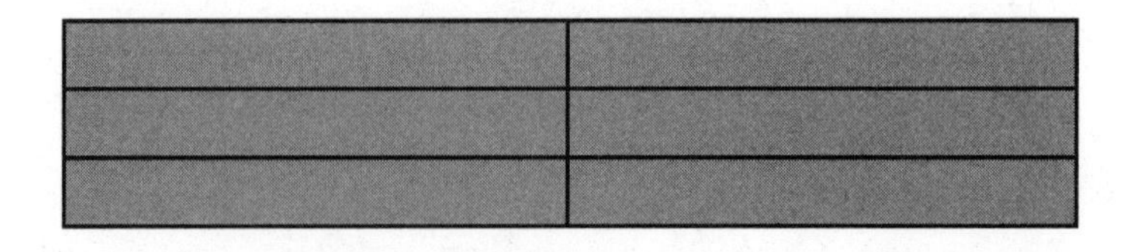

Here the *whole rectangle* is shaded. That's 1 rectangle. But it's also $\frac{6}{6}$, since 6 out of 6 pieces are shaded. Therefore, $\frac{6}{6} = 1$. In fact, it doesn't matter how many pieces you cut it into, all fractions with the same number in the numerator and denominator are equivalent to one whole, and therefore to each other. $\frac{1}{1} = \frac{2}{2} = \frac{6}{6} = \frac{15}{15} = \frac{99}{99} = 1$.

5. What about more than one whole?

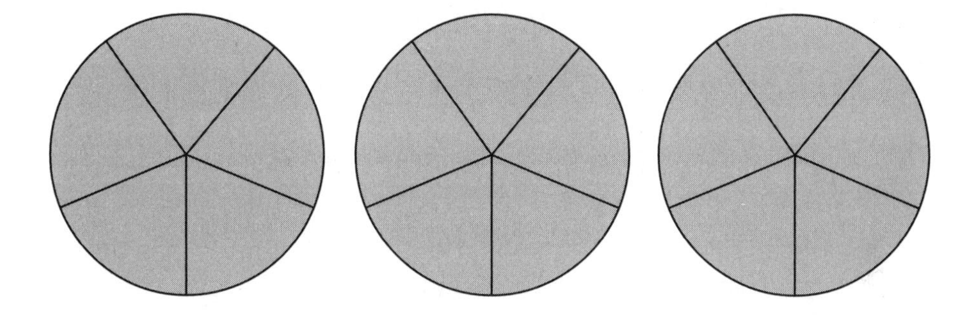

Here, 15 pieces are shaded, and each piece is $\frac{1}{5}$ of a whole. Therefore the total is $\frac{15}{5}$. But we have shaded 3 whole circles. So $\frac{15}{5} = 3$.

6. The following shows the relationships between tenths and hundredths, a relationship we will focus on in Chapter 3.

3 out of 10 rows is $\frac{3}{10}$. 30 out of 100 pieces is $\frac{30}{100}$. So $\frac{30}{100} = \frac{3}{10}$. $3 \times 10 = 30$, and $10 \times 10 = 100$.

SUMMARIZING THE PATTERN

By now you may have noticed a pattern. Two fractions are equivalent if the numerator and denominator of one fraction can both be multiplied by the *same number* to make the other fraction.

7. $\frac{8}{15}$ is equivalent to $\frac{32}{60}$ because $8 \times 4 = 32$ and $15 \times 4 = 60$. If you don't believe it, go ahead and draw a picture!

8. Go back to Example #5 for a moment. In the fraction $\frac{15}{5}$, we could *divide* both the numerator and the denominator by 5, giving us $\frac{3}{1}$. A denominator of 1 tells us that the whole has been cut into 1 equal-sized piece—in other words, it hasn't been cut at all! Therefore, $\frac{3}{1} = 3$. **Any whole number may be expressed as a fraction with a denominator of 1.**

We will spend more time discussing equivalent fractions later in the chapter.

Adding and Subtracting Like Fractions

You will sometimes see the terms like fractions and unlike fractions. Two fractions are considered **like fractions** if they have the same denominator. Since their denominators are the same, like fractions refer to pieces of the same overall size, which will make them easy to add and subtract. **Unlike fractions** have different denominators and refer to pieces of different overall sizes. Adding and subtracting unlike fractions can get tricky, so we begin by adding like fractions.

Examples

1. What is $\frac{1}{5} + \frac{1}{5}$? You have two pieces that are each $\frac{1}{5}$ of the whole—that's $\frac{2}{5}$. It's as simple as shading: $\frac{1}{5} + \frac{1}{5} = \frac{2}{5}$.

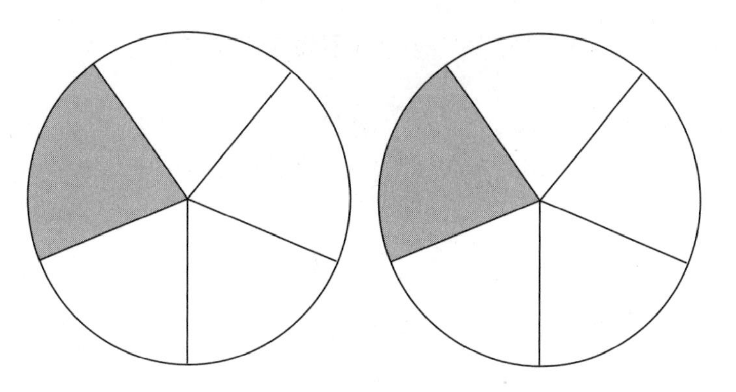

This diagram shows the 2 pieces as parts of different wholes. The next example shows the parts in the same whole.

2. $\frac{2}{9} + \frac{4}{9} = \frac{6}{9}$

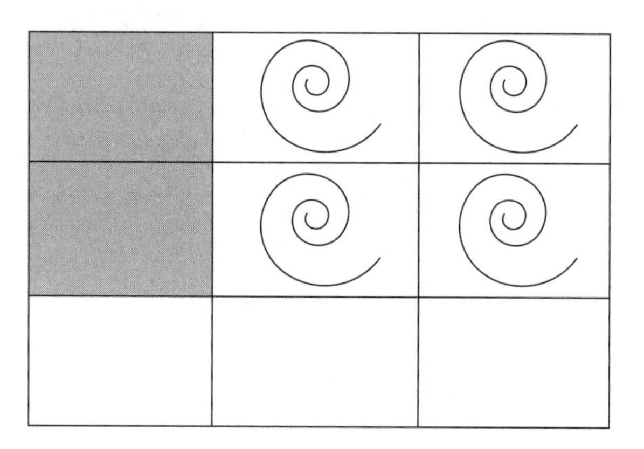

$\frac{2}{9}$ of the rectangle is shaded, and $\frac{4}{9}$ has spirals. Altogether, $\frac{6}{9}$ of the rectangle is filled in in some way.

3. Marco is reading a 250-page book. On Saturday, he read 29 pages, and on Sunday he read 22 pages. If he wants to know what *fraction* of the book he has read, we can think of it this way: $\frac{29}{250} + \frac{22}{250} = \frac{51}{250}$.

4. Lily had $\frac{3}{5}$ of a cookie that her mother gave her. If she gives $\frac{1}{5}$ to her brother Paul, she will have $\frac{2}{5}$ left: $\frac{3}{5} - \frac{1}{5} = \frac{2}{5}$.

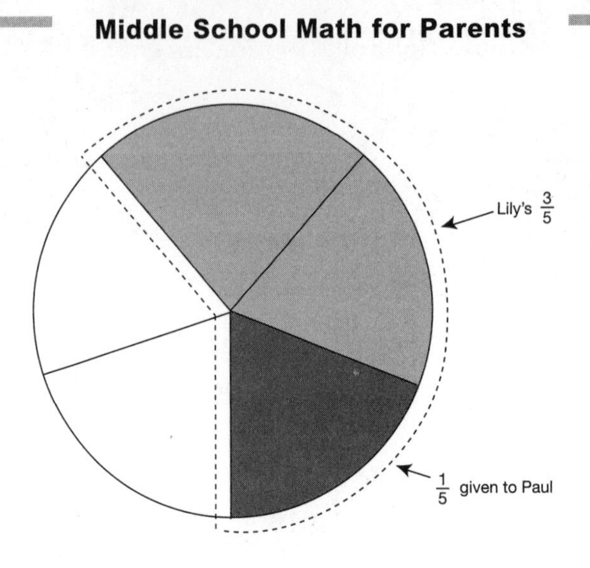

MAKING SENSE OF PROCEDURES

When adding or subtracting like fractions, simply add or subtract the numerators because you are simply adding and subtracting equal pieces. The denominator does not change because all the pieces are the same size.

Adding and Subtracting Unlike Fractions

Now that you understand how to add and subtract like fractions, the next step in our gentle approach to fractions is adding and subtracting *unlike* fractions. Before we delve into how to do it, let's examine why the method we have for like fractions does not work.

You know the old expression "apples and oranges"? If you are shopping and you pick up 4 apples and 3 oranges, how do you think about adding them together?

$3 + 4 = 7$, right? But 7 what? You don't have 7 apples, and you don't have 7 oranges. You have 7 fruits. Adding groups of *different* things cannot be done without changing the way you describe the things.

Suppose you have two pizzas, both the same size. If both pizzas are cut into the same number of equal slices, you can easily add slices:

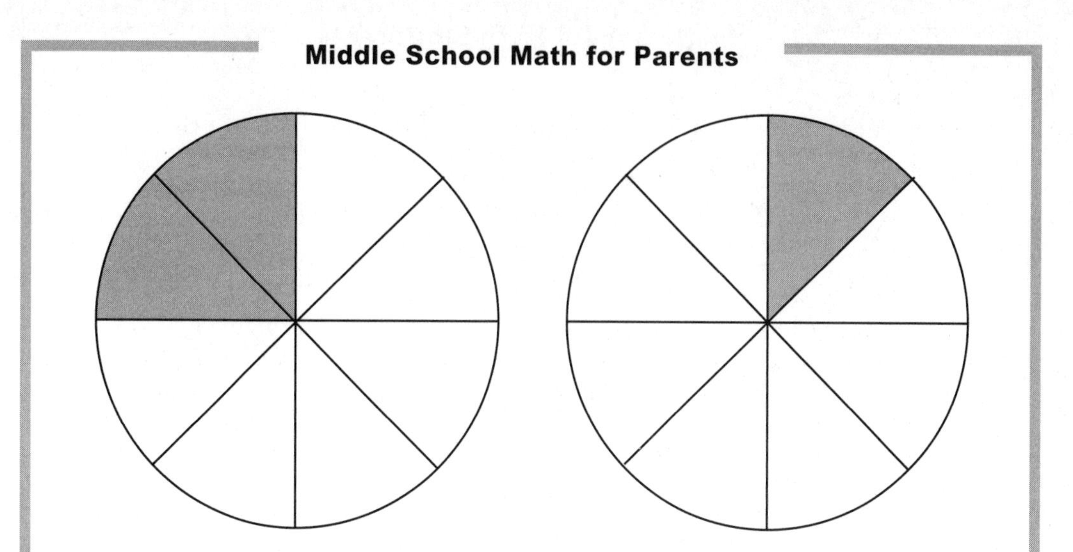

If you eat 2 slices of one pizza and 1 slice of the other, you have eaten $\frac{3}{8}$ of a pizza, since each slice is $\frac{1}{8}$ of the pizza. But what if the pizzas are not cut the same way?

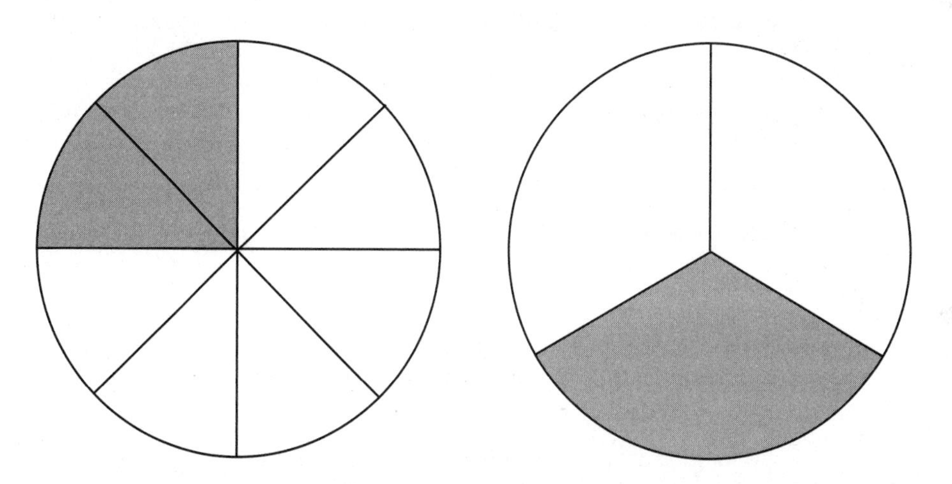

If you have eaten the shaded slices here, you could still say you ate 3 *slices*, but you did not eat $\frac{3}{8}$ of a pizza. You ate $\frac{2}{8}$ of one pizza and $\frac{1}{3}$ of another pizza. What does that add up to? It's not what you may think.

Adding and subtracting fractions is simple when the denominators are the same. But when the denominators are different, a more sophisticated method is needed. In order to add apples and oranges, we must generalize and call them both "fruit." In order to add thirds and fourths, we need to use equivalent fractions to give them all a common unit, or denominator.

We saw earlier that a fraction can be turned into a different, equivalent form when you multiply the numerator and the denominator by the same number. Since we already know how to add fractions that have the same denominator, we can use the idea of equivalent fractions to turn any fraction addition problem into a like-denominators problem.

Suppose we need to add $\frac{1}{2} + \frac{1}{4}$. The denominators are different, but we can easily convert $\frac{1}{2}$ to an equivalent form with a denominator of 4. Since $2 \times 2 = 4$, we can multiply both the numerator and denominator by 2:

$$\frac{1 \times 2}{2 \times 2} = \frac{2}{4}$$

Since $\frac{1}{2} = \frac{2}{4}$, we can rewrite our addition problem as $\frac{2}{4} + \frac{1}{4}$, which we know how to do—it's $\frac{3}{4}$.

We can use this method any time one of the denominators is a multiple of the other. Here's another example: $\frac{2}{3} + \frac{5}{6}$.

Since $3 \times 2 = 6$, we can multiply the numerator and denominator of $\frac{2}{3}$ by 2:

$$\frac{2 \times 2}{3 \times 2} = \frac{4}{6}$$

So we are really adding $\frac{4}{6} + \frac{5}{6}$, which is $\frac{9}{6}$.

MAKING SENSE OF PROCEDURES

Adding and subtracting like fractions is simple. To add or subtract unlike fractions, use what you know about equivalent fractions to turn it into a like-fractions problem.

Sometimes you will need to add fractions where neither denominator is a multiple of the other, like $\frac{1}{2} + \frac{2}{5}$. In this case, to get a common denominator, we need to convert *both* fractions to an equivalent form. The first step is to decide what our common denominator should be. You may remember the phrase **least common denominator**. Many teachers make a big deal of having the common denominator be the smallest possible number that is a multiple of both denominators. In this case, that would be 10, since there is no smaller number that both 2 and 5 go into evenly. But it does not necessarily have to be the least common denominator. For this problem, a denominator of 20, 50, or 100 would work just as well.

Let's go with 10 just because it is convenient. To get both fractions into equivalent forms with a denominator of 10, we need to do some multiplication:

$$\frac{1 \times 5}{2 \times 5} = \frac{5}{10} \quad \text{and} \quad \frac{2 \times 2}{5 \times 2} = \frac{4}{10}$$

So $\frac{1}{2} + \frac{2}{5}$ is the same as $\frac{5}{10} + \frac{4}{10}$, which is $\frac{9}{10}$.

For another example, let's try $\frac{3}{4} + \frac{5}{6}$. For our common denominator, we could use 12 or 24 or 36, etc. Let's go with 24, just to see what happens when we don't use the least common denominator.

$$\frac{3 \times 6}{4 \times 6} = \frac{18}{24} \quad \text{and} \quad \frac{5 \times 4}{6 \times 4} = \frac{20}{24}$$

So we have $\frac{18}{24} + \frac{20}{24} = \frac{38}{24}$. The fraction $\frac{38}{24}$ is not in lowest terms because both the numerator and the denominator are divisible by 2. If we divide both by 2, we will get an equivalent fraction in lowest terms.

$$\frac{38 \div 2}{24 \div 2} = \frac{19}{12}$$

$\frac{3}{4} + \frac{5}{6} = \frac{19}{12}$, or $1\frac{7}{12}$

For practice, try solving the same problem with a common denominator of 12, instead of 24. See if you get the same answer!

REMEMBER TO VISUALIZE

When finding an equivalent fraction, we either multiply or divide both the numerator and denominator of the original fraction by the same number. It is very important to remember why this works. When we converted $\frac{3}{4}$ to $\frac{18}{24}$, we multiplied both the numerator and denominator by 6. If we think back to our models of fractions, the equivalence between $\frac{3}{4}$ and $\frac{18}{24}$ really means that 3 of 4 equal-sized pieces represent the same amount as 18 out of 24 equal-sized pieces. In pictures, we can see that if we have an object that is cut into 4 equal pieces, and we cut each of *those* pieces into 6 smaller equal-sized pieces, we will end up with a total of 24 pieces, 18 of which are equivalent to our original $\frac{3}{4}$ of the whole.

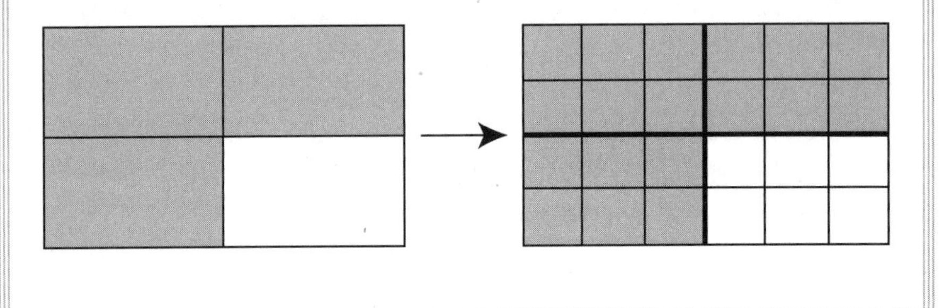

Multiplying a Fraction and a Whole Number

In Chapter 1, we covered two important ideas about multiplication, which we will continue to use throughout this book. The first is the idea that multiplication is really *repeated addition.* The second is the idea that multiplication is *commutative*—that is, when multiplying two numbers, it does not matter which number is written first—5 × 8 is the same as 8 × 5. If you have 5 groups of 8 objects or 8 groups of 5 objects, either way it's 40 objects. The same idea works when the numbers are fractions.

So suppose we need to multiply $\frac{1}{3} \times 5$. We can think of this as a repeated addition problem—adding $\frac{1}{3}$ to itself 5 times:

$$\frac{1}{3} \times 5 = \frac{1}{3} + \frac{1}{3} + \frac{1}{3} + \frac{1}{3} + \frac{1}{3}$$

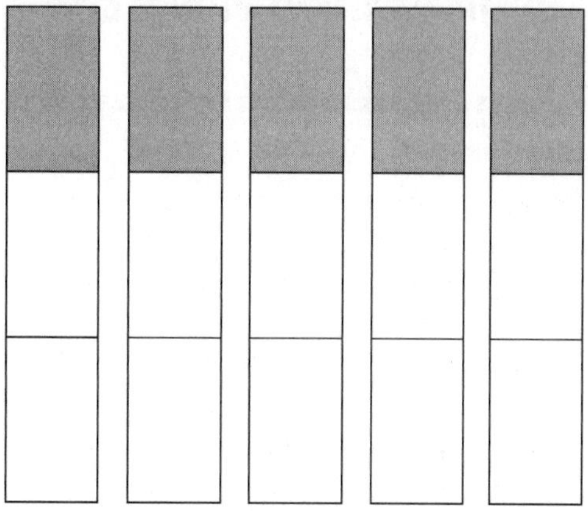

Since we are dealing with like denominators, we simply add all the numerators, which gives us $\frac{5}{3}$:

$$\frac{1}{3} \times 5 = \frac{5}{3}$$

Examples

1. What is $\frac{2}{7} \times 6$? We'll use repeated addition:

$$\frac{2}{7} \times 6 = \frac{2}{7} + \frac{2}{7} + \frac{2}{7} + \frac{2}{7} + \frac{2}{7} + \frac{2}{7}$$

$$\frac{2}{7} \times 6 = \frac{12}{7}$$

In mixed number form, this is $1\frac{5}{7}$.

2. What is $4 \times \frac{9}{10}$?

It doesn't matter that the whole number is now in front because of the Commutative Property. We can reverse the order of the factors or not—it amounts to the same thing.

$$4 \times \frac{9}{10} = \frac{9}{10} + \frac{9}{10} + \frac{9}{10} + \frac{9}{10}$$

$$4 \times \frac{9}{10} = \frac{36}{10}$$

We can convert this to mixed number form. $\frac{10}{10}$ makes one whole, so with $\frac{30}{10}$, we can make 3 wholes, with $\frac{6}{10}$ left over. That is, $3\frac{6}{10}$.

SUMMARIZING THE PATTERN

You may have noticed a pattern in the examples so far. If not, take a look at this summary:

- $\frac{1}{3} \times 5 = \frac{5}{3}$
- $\frac{2}{7} \times 6 = \frac{12}{7}$
- $4 \times \frac{9}{10} = \frac{36}{10}$

Do you see it? In each example, the final denominator has not changed from the original. To get the new numerator, we could simply multiply the original numerator by the whole number.

Remember that multiplication is commutative. This means that every multiplication problem can be thought of in two different ways. Take our first example, $\frac{1}{3} \times 5$. We could think of it as repeatedly adding $\frac{1}{3}$ to itself 5 times. Or we could think of it as repeatedly adding 5 to itself $\frac{1}{3}$ of a time. But wait—that is weird. What does it mean?

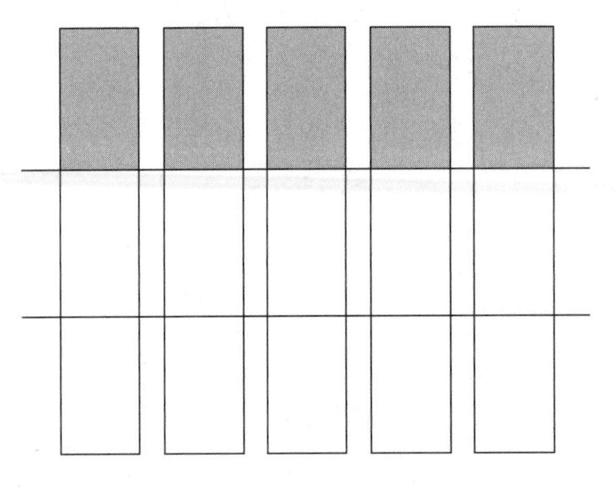

This diagram shows 5 whole bars, which, as a group, have been cut into equal thirds by the horizontal lines. $\frac{1}{3}$ of a group of 5 is the same as taking $\frac{1}{3}$ from *each* of the 5, and grouping them together.

THE LANGUAGE OF MATHEMATICS

When multiplying with fractions, use the word *of*. $\frac{5}{6} \times 12$ means $\frac{5}{6}$ *of* 12.

Multiplying Fractions Together

Now you understand how to multiply a fraction by a whole number. The next step in our gentle approach is to multiply two fractions together. When it comes to multiplying two fractions, it is very helpful to keep the word *of* in mind. A problem like $\frac{1}{2} \times \frac{2}{3}$ is really asking, "What is one-half *of* two-thirds?"

When we put it that way, it may become obvious to you. What is one-half of two-thirds? Half of two of anything is easy—it's one! There are two dogs at the park, and half the dogs are beagles—there's one beagle. There are two little boys in my family, and half of them still wear diapers—one boy still wears diapers. I had two-thirds of a pie, but I ate half of what I had—I ate $\frac{1}{3}$ of the pie.

For more complicated fraction problems, it is great to have a visual model. We can practice using our visual model on a simple problem like $\frac{1}{2} \times \frac{2}{3}$. We will use what is known as an area model. Here's $\frac{2}{3}$:

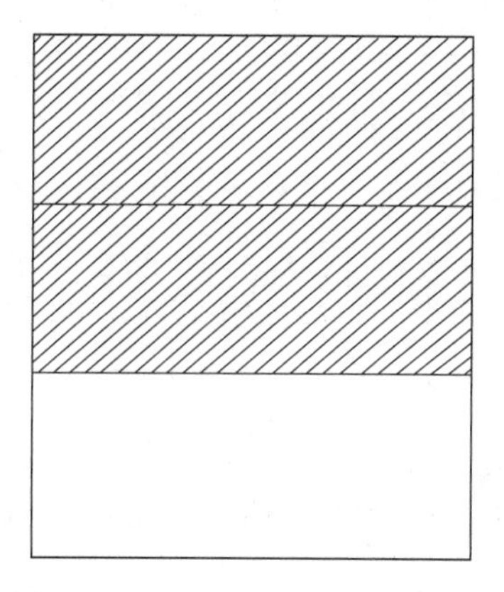

Now we cut it in half. The trick is to make a cut that is perpendicular to the old cuts. Since the thirds were separated horizontally, we will cut the picture in half vertically:

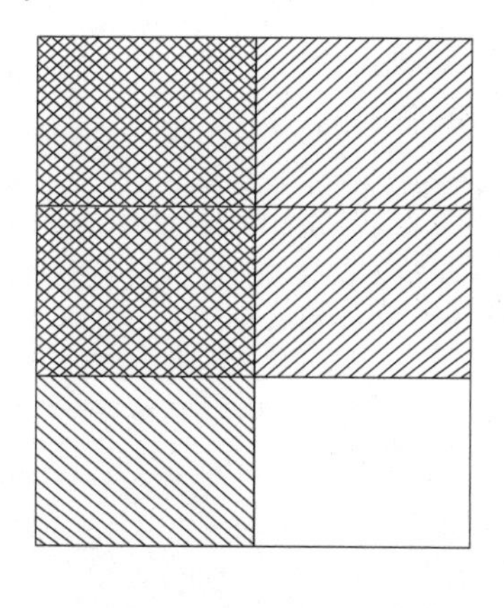

In this picture, we have cut the thirds in half, creating a total of 6 equal-sized pieces. We have shaded one half of each third on top of the shading we already had. Notice that now there are 2 pieces *double-shaded* out of 6. That is our product because those two pieces represent *half of two-thirds*. Therefore, $\frac{1}{2} \times \frac{2}{3} = \frac{2}{6}$, which is equivalent to $\frac{1}{3}$.

For a slightly more complicated example, let's try $\frac{3}{5} \times \frac{1}{4}$. Here's $\frac{3}{5}$:

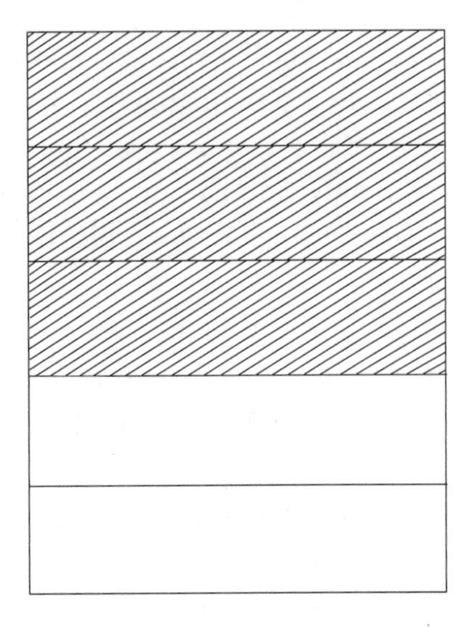

And now we cut each fifth into 4 equal pieces and shade $\frac{1}{4}$:

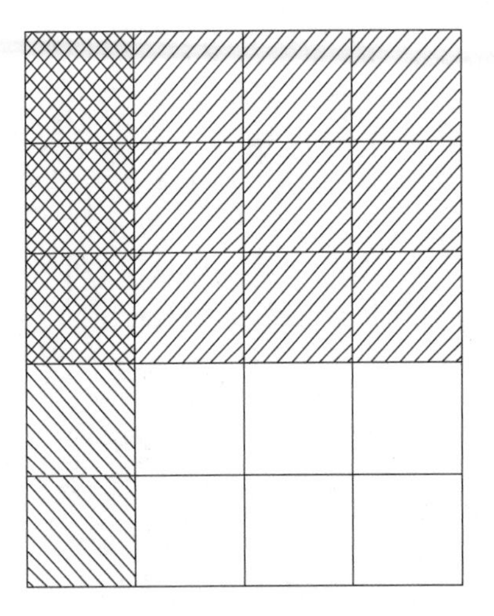

Altogether, 3 pieces are double-shaded, out of 20 total pieces. Therefore, $\frac{3}{5} \times \frac{1}{4} = \frac{3}{20}$.

MAKING SENSE OF PROCEDURES

You may remember learning to multiply the numerators together and multiply the denominators together. The visual models we are using justify why that procedure works. If we are multiplying $\frac{3}{5} \times \frac{2}{4}$, we are first cutting a whole into 5 equal pieces, then cutting each of those pieces into 4 equal pieces, resulting in 20 equal pieces altogether. Our denominator is $5 \times 4 = 20$. The numerators show that we are shading in 1 piece from each of 3 rows, so our numerator is $1 \times 3 = 3$.

The old rule still works, but students learn better when they learn the reasons for a rule, not just the rule itself. It is more important to understand why fraction multiplication works the way it does than to simply memorize how to do it.

Examples

For the following examples, try using an area model to show how the multiplication works.

1. $\frac{2}{5} \times \frac{1}{3} = ?$
2. $\frac{3}{8} \times \frac{1}{3} = ?$
3. $\frac{3}{10} \times \frac{5}{6} = ?$

Dividing with Unit Fractions

We've covered three of the basic operations, and now it's time for division. Division is usually the most difficult operation for students to grasp, whether it's with whole numbers, integers, fractions, or algebraic expressions. As we continue our gentle approach, we'll begin with just whole numbers and unit fractions.

Dividing a Whole Number by a Unit Fraction

Suppose, for example, we are asked to divide 6 by $\frac{1}{4}$. With division, it often helps to put the problem into words.

A division problem always asks one of two questions. The first is *How many times does one number go into another?* $35 \div 7$ is asking how many times 7 goes into 35, or how many groups of 7 are in 35. Since $7 \times 5 = 35$, there are 5 7s in 35. The other question that division asks is *How big would the groups be if we split one number into a certain number of equal groups?* $35 \div 7$ can be thought of in this way, too—if I have 35 tickets for a concert, and I'm splitting them into 7 equal groups, how many tickets would each group receive?

Our fraction example, $6 \div \frac{1}{4}$, is asking how many times $\frac{1}{4}$ goes into 6, or *How many fourths are in 6 wholes?*

We can answer this with a simple diagram:

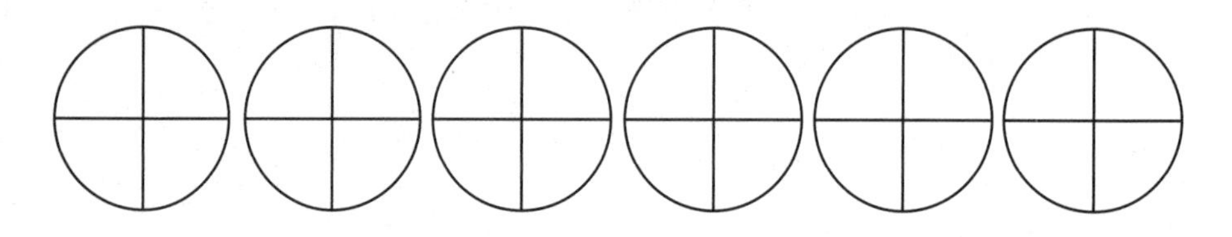

The diagram shows our 6 wholes, each cut into fourths. Each piece is $\frac{1}{4}$, and there are a total of 24 pieces. Therefore, $6 \div \frac{1}{4} = 24$.

Examples

1. $8 \div \frac{1}{2}$

In this diagram, 8 triangles have been cut in half. It takes a total of 16 pieces to make 8 triangles, so $8 \div \frac{1}{2} = 16$.

By now you may have noticed a pattern. If not, the next example will illustrate it.

2. $26 \div \frac{1}{9}$

For this one, let's *not* draw the picture. If you're inclined to draw 26 shapes and cut each one into 9 pieces, you may be missing the concept. Instead, let's simply visualize it. If there are 26 shapes with 9 pieces inside each one, then we have 26 groups of 9, which is a total of 26 × 9 = 234 pieces. Therefore, $26 \div \frac{1}{9} = 234$.

If you have noticed the pattern—*that dividing a whole number by a unit fraction is the same as multiplying the whole number by the denominator*—then you are in good shape for what's coming up. Again, try not to memorize this as a rule, but rather think of it as a helpful generalization of a pattern. This kind of thinking will go a long way toward helping you and your child see the beauty in math, because math is not about rules—it is about finding, understanding, and describing patterns.

Fractions Are Quotients

Remember earlier when we emphasized that a fraction is a number? Now we're going to dig a little deeper. A fraction is not just a number, it is a quotient—that is, every fraction is the answer to a division problem.

Examples

1. $\frac{1}{8}$ is the answer to the division problem 1 ÷ 8. That is, if you take 1 whole and divide it into 8 parts:

To put it in words, 1 ÷ 8 asks, *How many times does 8 go into 1?* This is confusing for many students, because how can a larger number go into a smaller number? We mentioned earlier that division can ask two different questions. For a problem like this, it is helpful to use the other question. How large would the "groups" be if we split 1 into 8 equal groups? In this way, it is easy to see that $1 \div 8 = \frac{1}{8}$.

2. $\frac{3}{5}$ is the answer to the division problem $3 \div 5$.

$3 \div 5$ asks how large the groups would be if we cut 3 wholes into 5 equal pieces.

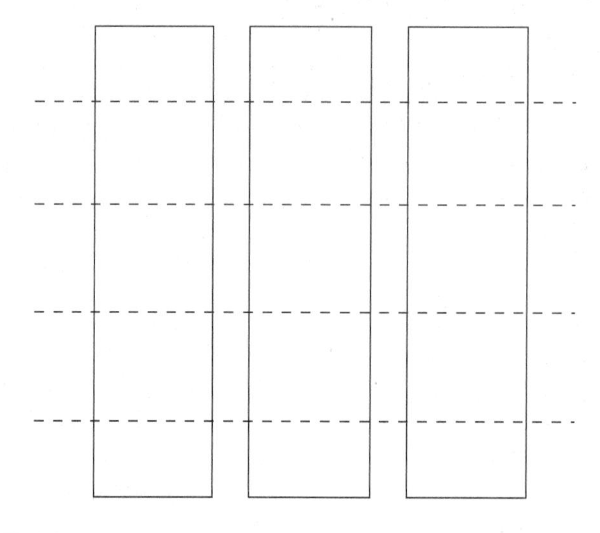

Here we have 3 wholes, cut into 5 equal pieces, or rows. Our quotient is the size of *one* of those rows, since division is asking the size of one group. Each of the 5 rows has 3 little pieces to it, and each little piece is $\frac{1}{5}$ of a whole. Therefore, $3 \div 5 = \frac{3}{5}$.

3. Of course, a whole number can be a quotient as well. For example, 3 is the quotient of $12 \div 4$. Let's think of this in terms of fractions.

$12 \div 4$ asks how large the groups will be when we divide 12 into 4 equal groups. We already know 3 goes into 12 a total of 4 whole times. But what happens if we put this division problem in a diagram, just like the previous two examples?

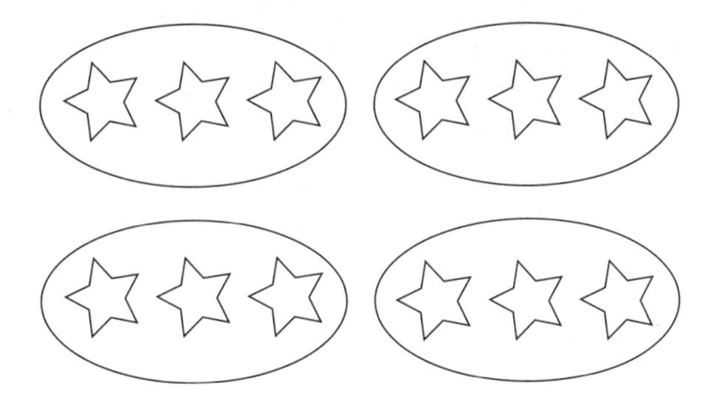

This diagram shows that there are 4 groups of 3 in 12. We can write the division problem 12 ÷ 4 as a fraction, $\frac{12}{4}$. Therefore, we have the idea that $\frac{12}{4} = 3$. *Sometimes a fraction can be equivalent to a whole number.*

REMEMBER

Any fraction $\frac{\textit{numerator}}{\textit{denominator}}$ is the quotient to the division problem *numerator ÷ denominator*.

This is why you will often see both fractions *and* division written with a slash. $\frac{1}{2}$ can mean one half *or* 1 divided by 2.

For reasons we will see shortly, this is possibly the most important takeaway from this chapter.

Dividing a Unit Fraction by a Whole Number

We have divided whole numbers by unit fractions by counting (and multiplying) to find how many times the fraction goes into the whole number. Now that we understand that fractions and division are two forms of the same thing, we can turn our attention to more complex division problems. Let's begin by dividing a unit fraction by a whole number.

For example, the problem $\frac{1}{3} \div 7$ asks how large the pieces are when we cut $\frac{1}{3}$ into 7 equal pieces.

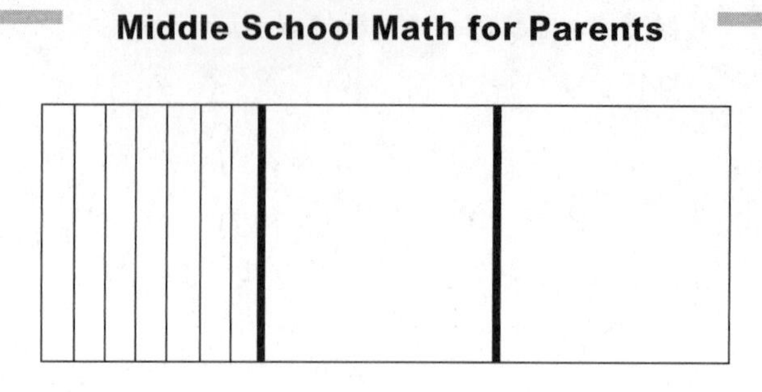

Here we have cut $\frac{1}{3}$ of a whole into 7 equal pieces. But we cannot tell what fraction one of those pieces is unless we think about how many will fit in the entire whole. We need to cut all three thirds:

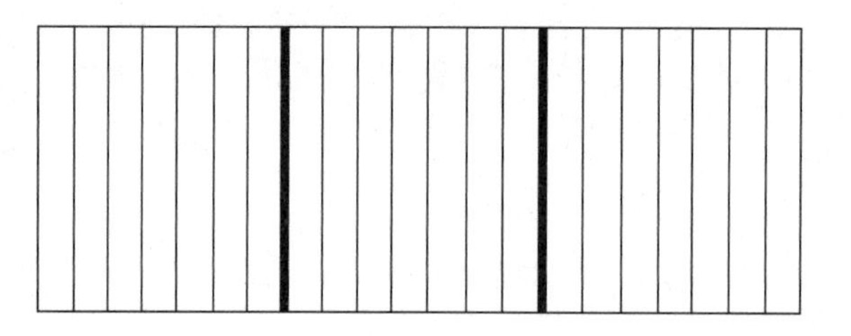

It's easy to see here that there are 21 equal-sized pieces. Therefore, cutting $\frac{1}{3}$ into 7 equal pieces gives us pieces that are each $\frac{1}{21}$ of a whole: $\frac{1}{3} \div 7 = \frac{1}{21}$.

Again, you may notice that multiplication is absolutely essential to solving a division problem. The diagrams should help you understand why. When dividing with fractions, multiplication is necessary because to answer the division problem, we often need to cut the wholes into smaller pieces.

Dividing by Expressing Fractions as Quotients

There is another way to think about a problem like $\frac{1}{3} \div 7$, using the idea that a fraction is a quotient. If $\frac{1}{3} = 1 \div 3$, we can rewrite the division problem in the following way:

$$\frac{1}{3} \div 7 = 1 \div 3 \div 7$$

In words, this is asking us to divide 1 first by 3, and then by 7. This may be easier to conceptualize than the previous method. Here is a sheet cake (1 whole), cut into 3 equal pieces:

And now we cut each of the pieces into 7 pieces:

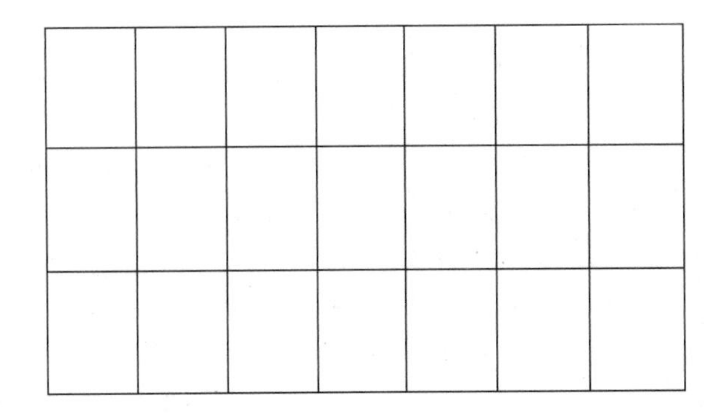

We have 21 equal pieces.

Example

1. $\frac{1}{4} \div 5$

Let's cut fourths into 5 equal pieces:

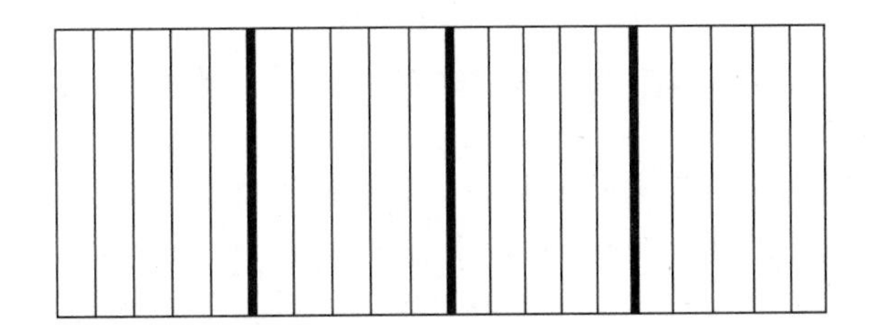

Since the whole now contains 20 equal-sized pieces, $\frac{1}{4} \div 5 = \frac{1}{20}$.

$1 \div 4 \div 5$ is the same as dividing 1 whole into 20 pieces.

Let's make a fact family out of this problem, just to see how these three numbers are related:

$5 \times \frac{1}{20} = \frac{1}{4}$	If you add $\frac{1}{20}$ to itself 5 times, it will make $\frac{1}{4}$.
$\frac{1}{20} \times 5 = \frac{1}{4}$	$\frac{1}{20}$ of 5 is $\frac{1}{4}$; if you cut 5 wholes into a total of 20 equal pieces, each piece is $\frac{1}{4}$ of one whole.
$\frac{1}{4} \div 5 = \frac{1}{20}$	If you cut $\frac{1}{4}$ into 5 equal pieces, each piece will be $\frac{1}{20}$ of one whole; $\frac{1}{20}$ of 5 goes into $\frac{1}{4}$.
$\frac{1}{4} \div \frac{1}{20} = 5$	$\frac{1}{20}$ of a whole goes into $\frac{1}{4}$ of a whole 5 times.

Dividing with *Any* Fractions

So far, we have done division with unit fractions only. Now we turn our attention to more general problems, where the numerator can be any whole number. You may remember the old algorithm of multiplying by the reciprocal of the second number. While this certainly works, we again come up against the classic pitfall involved in a procedural approach. Math teachers often joke, "Ours is not to reason why, we just flip and multiply," because division with fractions is one of the most difficult concepts to understand in all of K-12 math, and even some teachers who believe strongly in a conceptual approach fall back on the old rule here.

As we have done throughout this chapter, we will use a gentle approach in small steps for the comfort of the fractionphobic. Instead of using the old rule, we will cover three different methods that will help you and your child make sense of division with fractions.

First Method: Building on Number Sense

Suppose we are asked to solve $4 \div \frac{2}{3}$. So far, we do not have a method for solving this. But we do know how to solve $4 \div \frac{1}{3}$: Since $\frac{1}{3}$ goes into 1 whole 3 times, and there are 4 wholes in 4, we simply multiply 3×4. $4 \div \frac{1}{3} = 12$. Now let's relate this simplified problem to our original problem, $4 \div \frac{2}{3}$. Mathematicians often use this strategy, taking a difficult problem and connecting it to an easier problem that has some similarities. In both of these cases, we are dividing 4 by a fractional divisor. In our original problem, the fraction is twice as much as the fraction in the simplified version of the problem.

What happens in other division problems when you double the divisor?

$$20 \div 5 = 4$$

$$20 \div 10 = 2$$

When we doubled the divisor (turning 5 into 10), the quotient became half as much. The same thing happens every time.

$$60 \div 6 = 10$$

$$60 \div 12 = 5$$

Why does this happen? Because dividing 60 by 12 is the same as dividing it first by 6 and then by 2, since $12 = 6 \times 2$.

$$60 \div 12 = (60 \div 6) \div 2$$

The same is true when our divisor is a fraction. Therefore, $4 \div \frac{2}{3}$ is the same as dividing 4 first by $\frac{1}{3}$, then by 2:

$$4 \div \tfrac{2}{3} = (4 \div \tfrac{1}{3}) \div 2$$

We know that $4 \div \frac{1}{3} = 12$, so $4 \div \frac{2}{3}$ is half as much, or 6. What we have done, essentially, is multiply 4 by the denominator, then divide by the numerator—in other words, "flip and multiply."

Second Method: Visual Models

Now let's use a visual model to see why $4 \div \frac{2}{3} = 6$. One model that we have not used yet is a number line, which is very helpful for fraction division.

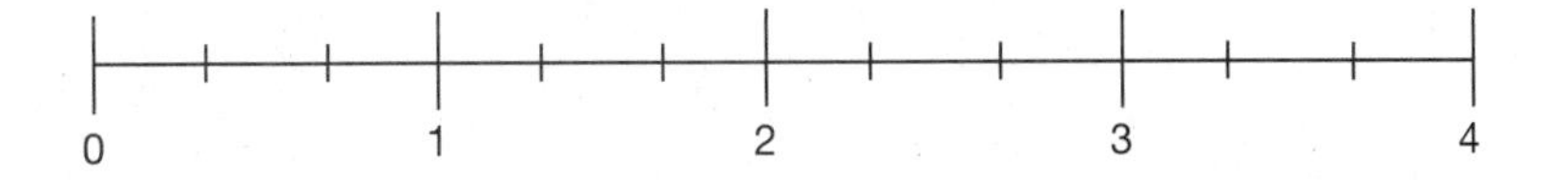

In this diagram, we have drawn a number line from 0 to 4, and we have divided each whole into equal thirds. We are trying to find out how many times $\frac{2}{3}$ goes into 4. All we need to do is simply circle groups of $\frac{2}{3}$:

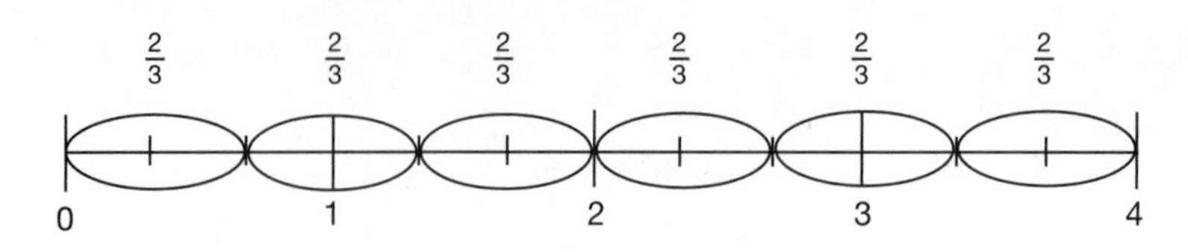

Now it's quite clear to see that there are 6 groups of $\frac{2}{3}$ in 4 wholes.

What's great about the number line model is that it works even when the divisor does not go in evenly. Suppose that instead of dividing 4 by $\frac{2}{3}$, we had to divide 5. In this case, we cannot complete the last group of $\frac{2}{3}$:

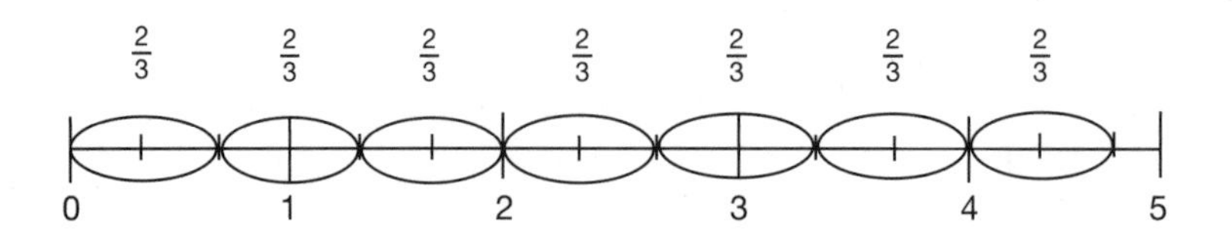

We have 7 groups of $\frac{2}{3}$, but at the end, we have $\frac{1}{3}$ left over. Think about the question: *How many times does $\frac{2}{3}$ go into 5?* It goes in a little more than 7 times. But how much more? We want to cram another $\frac{2}{3}$ in there, but there's only room for $\frac{1}{3}$, which is *half as much room as we need*. Therefore, we could put *half of another* $\frac{2}{3}$ in 5. In other words, $\frac{2}{3}$ goes into 5 wholes 7 and $\frac{1}{2}$ times:

$$5 \div \frac{2}{3} = 7\frac{1}{2}$$

Your child will see lots of word problems that require division with fractions. For example, suppose you are baking cookies for a party. The recipe for one batch of cookies calls for $\frac{3}{4}$ cups of sugar. You have $2\frac{1}{2}$ cups of sugar to use. How many batches can you make if you use all the sugar?

This problem is asking the question *How many times does $\frac{3}{4}$ go into $2\frac{1}{2}$?* To figure it out, let's use a number line model. Here's our $2\frac{1}{2}$ cups of sugar, cut into half-cup intervals:

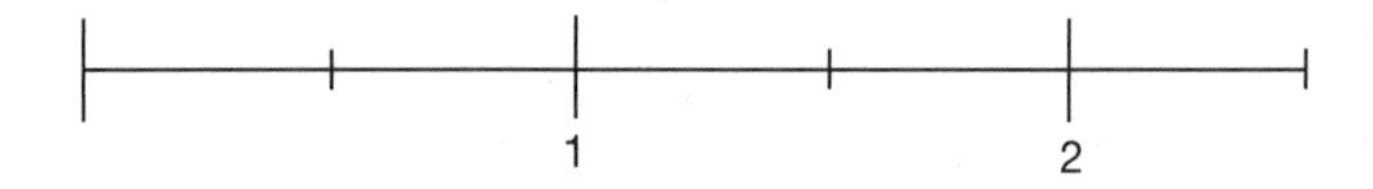

Since we are trying to divide it up into pieces that are $\frac{3}{4}$ cups in size, we need to subdivide our number line into fourths, so we can group 3 at a time:

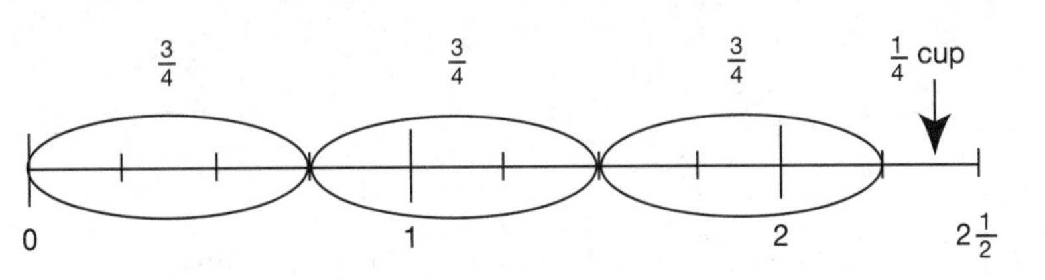

Now we can see that we have enough for 3 full batches, plus $\frac{1}{4}$ cup of sugar left over. If we want to use that $\frac{1}{4}$ cup as well, we need to figure out what fraction of a batch that will make. Since a full batch requires $\frac{3}{4}$ cups, $\frac{1}{4}$ of a cup will make $\frac{1}{3}$ of a batch. If that doesn't make sense, take a moment and think about it:

$$\frac{3}{4} \text{ cup} = 1 \text{ batch}$$

$$\frac{1}{4} \text{ cup} = \frac{1}{3} \text{ batch}$$

Therefore, you could make $3\frac{1}{3}$ batches of cookies:

$$2\frac{1}{2} \div \frac{3}{4} = 3\frac{1}{3}$$

Examples

Let's try a couple more problems, using both of the methods we have covered so far.

1. $3\frac{2}{3} \div \frac{5}{6}$

First, let's think about how many times $\frac{1}{6}$ goes into $3\frac{2}{3}$. Each whole contains $\frac{6}{6}$, and each $\frac{1}{3}$ is equivalent to $\frac{2}{6}$. Therefore:

$3 = \frac{18}{6}$

and $\frac{2}{3} = \frac{4}{6}$

So $3\frac{2}{3} = \frac{22}{6}$ (18 + 4 = 22), and therefore, $3\frac{2}{3} \div \frac{1}{6} = 22$.

But we are trying to divide by $\frac{5}{6}$, not $\frac{1}{6}$. As we saw earlier, multiplying the divisor by 5 means the quotient will be *divided* by 5. Therefore, $3\frac{2}{3} \div \frac{5}{6} = \frac{22}{5}$. In mixed number form, this is $4\frac{2}{5}$, since there are 20 fifths in 4 wholes.

Let's also use the number line method for this problem. We need to cut out $3\frac{2}{3}$ into $\frac{1}{6}$-sized pieces:

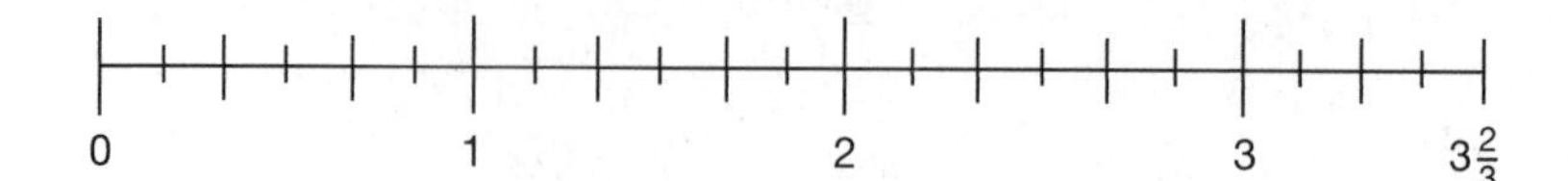

Here we have cut each whole into $\frac{6}{6}$, and each third into $\frac{2}{6}$. You can count the $\frac{22}{6}$ in this diagram. Now we group them in groups of $\frac{5}{6}$:

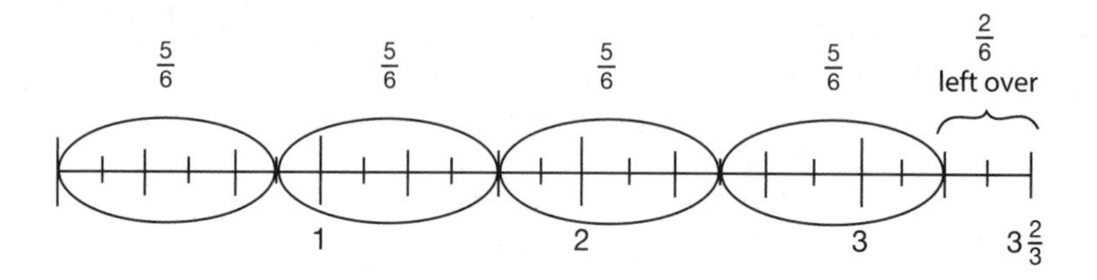

We have 4 groups of $\frac{5}{6}$ with $\frac{2}{6}$ left over. We need $\frac{5}{6}$ to make another whole group, so we have 2 out of 5, or $\frac{2}{5}$ of the next group. Therefore, $3\frac{2}{3} \div \frac{5}{6} = 4\frac{2}{5}$.

Third Method: Common Denominators

The number line method has proven to be very successful so far. But the problems we have looked at were chosen carefully for that method. In each example we have seen, the denominator of the divisor was a multiple of the denominator of the dividend. This means that when you draw the dividend on a number line, it is easy to cut each piece to get the divisor.

When the divisor and the dividend are not related in this way, another method is needed.

As an example, let's find the quotient of $\frac{3}{5} \div \frac{1}{4}$. If we draw these fractions on number lines, you can see that they do not fit together in any convenient way:

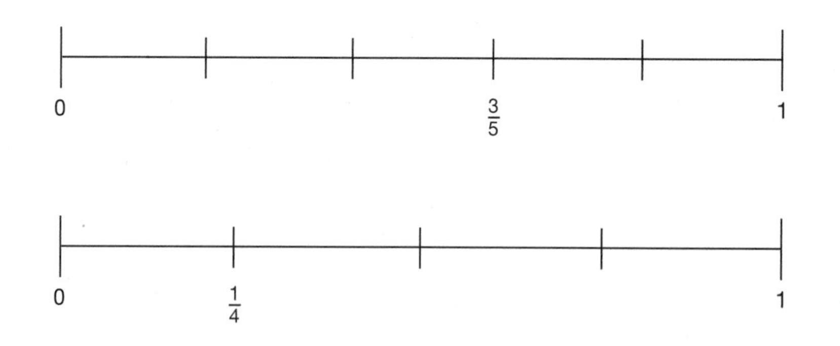

Since the pieces do not line up together, we need a different way to break them down. When we were dealing with addition and subtraction, we solved this problem by using a **common denominator**. The same thing works here, in this case with a denominator of 20. We can rewrite $\frac{3}{5} \div \frac{1}{4}$ as $\frac{12}{20} \div \frac{5}{20}$, which is easy enough to show on a number line:

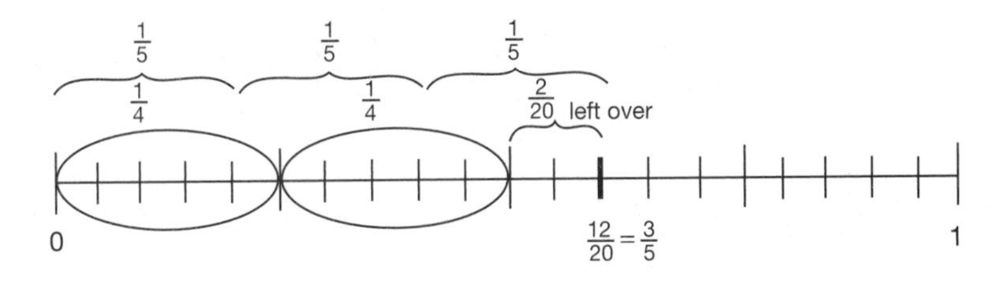

By dividing the whole into 20 equal pieces, we can find groups of 5, which are $\frac{1}{4}$ each. So we see that we have 2 groups, and $\frac{2}{20}$ left over—that is, 2 out of 5 toward the next group. Therefore, $\frac{12}{20} \div \frac{5}{20} = 2\frac{2}{5}$. In improper fraction form, $2\frac{2}{5} = \frac{12}{5}$, which is the quotient of the numerators! In other words, once you get a common denominator for a division problem, you can simply divide the numerators.

REMEMBER HOW AND WHY

If you can convert a division problem to an equivalent problem with common denominators, then your quotient is simply the quotient of the numerators. This works because after converting, you are dividing pieces of a certain size into groups of pieces that are the same size. At that point, it is no different than simple, familiar whole-number division.

Example

Ella needs $\frac{2}{3}$ of a cup of blueberries to make a batch of muffin batter. She has $1\frac{1}{4}$ cup of blueberries left, and she wants to use all of the blueberries. How many batches of batter can she make?

We need to divide: $1\frac{1}{4} \div \frac{2}{3}$. We can solve this problem by using a common denominator. 3 and 4 both go into 12, so let's convert both numbers to fractions with denominators of 12. For $1\frac{1}{4}$, this means converting to an improper fraction.

$$1\frac{1}{4} = \frac{5}{4} = \frac{15}{12}$$

$$\frac{2}{3} = \frac{8}{12}$$

So we are dividing $\frac{15}{12} \div \frac{8}{12}$. We already know we can divide the numerators because the units are all the same size (common denominator), but we'll draw a number line just to be sure:

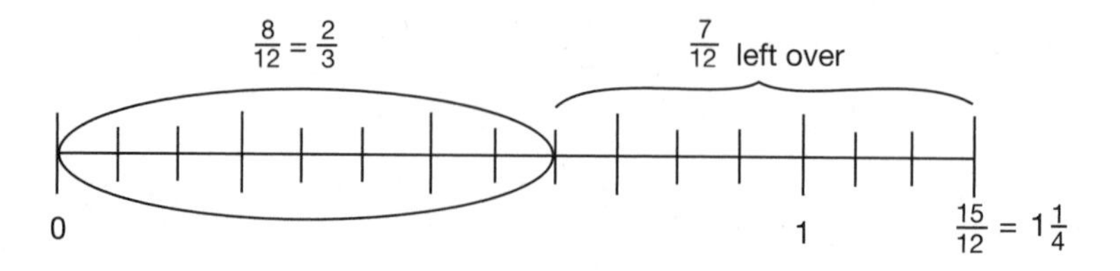

Here we have 1 batch and an extra $\frac{7}{12}$ cup of blueberries left over. How much batter can $\frac{7}{12}$ blueberries make? We need $\frac{8}{12}$ to make a full batch, so we could make $\frac{7}{8}$ of a batch. Therefore, we can make a total of $1\frac{7}{8}$ batches. As an improper fraction, that would be $\frac{15}{8}$.

Dividing Fractions: Summary

The three methods we have presented for dividing fractions—using number sense to relate to whole-number division, drawing number lines, and finding common denominators—are meant to help you and your child understand fractions and division of fractions on a conceptual level. As your child works through more and more problems, she will likely come up with shortcuts and may eventually master the "flip and multiply" method. When she does, it will mean much more to her than it ever did to those of us who were taught that method outright because she will have experienced the process of discovering it—that is, she will always understand where it comes from and why it works.

Summary

In this chapter, we have examined fractions, one of the major catalysts for the development of mathphobia for students. We know that your child will need to master fractions for future success in school and work, so we have taken a gentle approach that is intended to reduce that fear. By building the concepts slowly and with plenty of visual representations, we have made more sense of fractions and how they work, so that you and your child will have less reason to feel intimidated.

As we did in Chapter 1, we have approached the math in a way that emphasizes sense-making, visual representations, and justification, rather than rules that can be hard to remember and apply. Remember these important points:

- A fraction is a number—don't be afraid to treat it like one.
- Adding or subtracting like fractions is simple. Use a visual if you forget how.
- Adding or subtracting unlike fractions is difficult but can always be turned into a like-fractions problem.
- Multiplication is repeated addition.
- You can always use an area model, number line, or another visual representation when multiplying or dividing fractions.
- Dividing with unit fractions and whole numbers is simple!
- A complex problem, or any new kind of problem, can usually be related to a familiar or simpler problem, which can often save the day.
- We have three ways to divide fractions: (1) Relating to whole-number division or division with unit fractions; (2) Drawing number line models; and (3) Using common denominators.

In Chapter 3, we will focus on the base-ten system, specifically for decimals, which is another way of writing fractions. If your child can master place value, then he or she will be able to perform many calculations mentally without needing a calculator—or even paper and pencil.

Mental Math Rules!

Unlocking the Secrets of the Base-Ten System and Understanding Decimals

Why Mental Math Matters

Not too long ago, it used to be true that only engineers and other nerds carried pocket calculators. Today's teens are too young to remember that time. They have grown up in a world where everybody always has a calculator on them, and many students have come to rely on technology so heavily that they have lost sight of what math is all about.

Two problems arise from over-dependence on calculators.

The First Problem

Students who rely on calculators too often fail to think. When asked to explain an answer to a problem, they will say, "I did it on my calculator." There is a difference between *math* and *arithmetic*. You can think of arithmetic as all the stuff the calculator can do—the four basic operations.

But that is not mathematics. Mathematics is a structured system of organized thinking—math is reasoning and problem solving. Math is a way of describing the world by recognizing its many patterns and relationships. Arithmetic is a tool that is often used when solving mathematical problems, but math goes so much deeper than simple operations. So many people—students and adults alike—fail to understand this distinction.

Math is how people build bridges, predict the weather, create vaccines, and invent better cell phones. No calculator can do those things because they require human ingenuity, creative thinking, and problem solving. How did NASA put people on the moon? It was not as simple as calculating where the moon would be and making sure the Apollo ships were on the correct trajectory—though that alone was a challenging calculation. They had to consider every possible factor that could affect the mission, all the conditions that could endanger the astronauts. They had to invent all the items that would be needed on board the ship, then find a way to make everything fit into a small, light, aerodynamic vessel. Math is about the big picture, not the small but significant calculations along the way.

Students who think math is what a calculator can do are missing the point. Math is not about getting a right answer, it is about thinking—something that no machine can do as well as a human.

The Second Problem

Students who rely on calculators too much and too early fail to develop adequate skills with our number system. Calculators and other technology are incredible tools that *every* mathematician relies on at times. The trick is to make smart decisions about when a calculator is necessary and when it isn't. Some middle school students will reach for a calculator when a problem calls for them to add something as simple as 10 + 10. Of course in this extreme (but very real) example, which I have seen far too many times in my 8th grade classroom, it should be obvious that you can calculate the sum in your head faster than you can turn the machine on. But the real consequences of this reaction to arithmetic are far deeper and more problematic than time efficiency.

For example, suppose Kyle had frequent access to a calculator when learning multiplication and division around 3rd and 4th grades. In this case, he probably did not memorize the basic multiplication facts. He may have been able to recall them long enough for the quizzes, but if he was using a calculator for his homework, then there's a good chance they didn't stick with him for long.

Without having memorized the rows and columns of multiplication charts, he then would have had significant difficulty understanding the concept of equivalent fractions in 5th grade, and may never have fully grasped how to simplify fractions, leading to fractionphobia, as we discussed in Chapter 2.

But it gets worse when he gets to algebra, because now he's being asked to find factors and multiples of algebraic expressions like $4x^3y^5$ or to simplify fractions like $\frac{3x^2 - 12x}{9x^4}$. If he doesn't know what the factors of 24 are, how will he find the factors of $x^2 - 8x + 24$? These are not problems a cell phone calculator is going to help you with, and to be able to even enter such a problem into a more sophisticated computer algebra system or graphing calculator takes a certain amount of knowledge, which just isn't likely to be there for poor Kyle.

This story shows pretty plainly how technology can create a deficiency. For many students, this deficiency has to do with multiplication and division. For others, it comes in the form of a basic misunderstanding of the base-ten system—the foundation on which all of our numbers are built. Students who do not properly understand the base-ten system simply do not understand how numbers work.

What Is the Base-Ten System?

The **base-ten system** is one of humankind's most brilliant inventions. Our English alphabet has 26 letters, from which we can create an incredible variety of words. Our number system has 10 digits, from which we can create and represent *infinite* numbers by using place value.

Place value is a way of organizing numbers by their size. For example, the number 56 is made up of 5 *tens* and 6 *ones*. The place value of the 5 is tens, and the place value of the 6 is ones. As you move to the left from one digit to the next, the place value of each digit is ten times larger than the digit to its right. In a number like 222, the first 2 is ten times as much as the second 2, which is ten times as much as the third 2. Not only is this system very convenient for writing very large numbers (and very small ones, using decimals), but it also makes calculations very simple by allowing us to break numbers into their place value parts.

TERMS TO REMEMBER

One method that elementary and middle school teachers now use to help students understand place value is called **expanded form**. In expanded form, a number is broken up to show exactly how much place value each digit has. For example, in expanded form, the number 78,016 would be written as:

$$70{,}000 + 8{,}000 + 10 + 6$$

In this way, students learn that the 7 is not really a 7, but is 70,000, or 7 *ten thousands*. The 1 is not 1, but 1 *ten*. We could expand the number in a different way to show even more detail:

$$7 \times 10{,}000 + 8 \times 1{,}000 + 0 \times 100 + 1 \times 10 + 6 \times 1$$

This expansion makes it very clear that we are dealing with **powers of ten**—the numbers 1, 10, 100, etc., which make up our number system.

Note that while we often refer to zeros as place holders, in fact they are more than that. In the number 78,016, the zero is not just holding a place for some reason. It is there to tell us that we don't have any hundreds in this number, just 7 ten thousands, 8 thousands, 1 ten, and 6 ones.

Adding and Subtracting in Base-Ten

In the early grades, your child likely learned the classic addition algorithm for multi-digit whole numbers:

$$\begin{array}{r} {}^{1}472 \\ +\ 156 \\ \hline 628 \end{array}$$

Let's dissect that algorithm by breaking the numbers down into their place value parts.

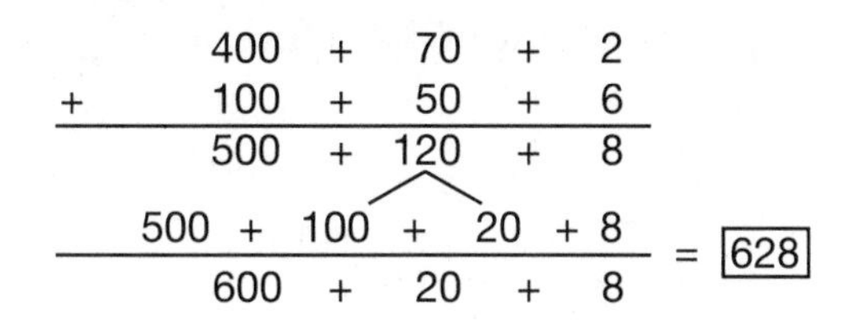

In other words, we add the hundreds, the tens, and the ones separately, then put them all together to find the sum. Notice that when adding the tens, we ended up with more than 100: 70 + 50 = 120. 7 tens plus 5 tens gave us 1 hundred and 2 tens, so we add the 1 hundred to our other hundreds. This is what "carrying the one" is all about, which many students don't understand. When you carry the one, you are really grouping a ten or a hundred, etc., with other numbers that have the same place value.

Subtraction works in much the same way. Let's consider the problem 371 – 135. Here is the classic algorithm, with borrowing:

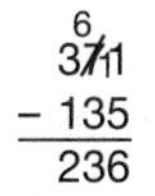

And here it is again, written in a way that shows how we are using place value to "borrow." 7 tens and 1 one is the same as 6 tens and 11 ones. By thinking of the number in this way, you can see why we can subtract the ones:

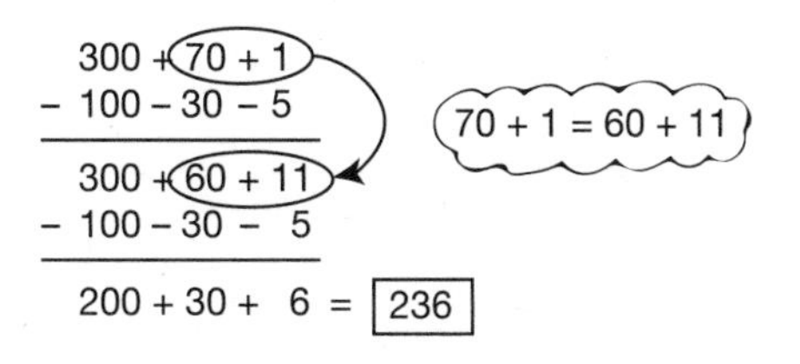

This second method may look like it takes a lot more work because it takes more space to write it out. But the point of this method is to help your child develop what math teachers call *fluency*—the ability to perform calculations efficiently, accurately, and often mentally. Students who only learn the "classic" algorithms for the four operations may never be able to perform those operations without paper and a pencil because their method does not make sense to them. If your child practices adding and subtracting whole numbers by using expanded form, he will likely become fluent much more quickly—and that means when decimals come along in 6th grade, he will be ready.

Examples

1. $384 + 819$

$$\begin{array}{r} 384 \\ +\ 819 \\ \hline \end{array} = \begin{array}{rcrcr} 300 & + & 80 & + & 4 \\ +\ 800 & + & 10 & + & 9 \\ \hline 1{,}100 & + & 90 & + & 13 \end{array}$$

$$1{,}000 + 100 + 90 + 10 + 3$$

$$1{,}000 + 100 + 100 + 3$$

$$1{,}000 + 200 + 3$$

$$\boxed{1{,}203}$$

In this case, when we add the place values, a few tricky things happen. The hundreds add up to 1,100—one thousand and one hundred. The ones add up to 13, one ten and 3 ones. I have shown baby steps in the process to solving this. Take a few moments to see if you understand what was done in each step. You normally would not write down each one of these steps—what I call "baby steps"—because with practice you will find that several steps can be done quickly in your head.

2. $463 - 295$

$$\begin{array}{r} 463 \\ -\ 295 \\ \hline \end{array} = \begin{array}{rcrcr} 400 & + & 60 & + & 3 \\ -\ 200 & - & 90 & - & 5 \\ \hline 400 & + & 50 & + & 13 \\ -\ 200 & - & 90 & - & 5 \\ \hline 300 & + & 150 & + & 13 \\ -\ 200 & - & 90 & - & 5 \\ \hline 100 & + & 60 & + & 8 \\ \hline \end{array}$$

$$\boxed{168}$$

Again, take a moment to make sure you can follow each of these steps. Then, try making up some more problems for you and your child to try in your heads. Start with two-digit numbers, and build up to three- or four-digit numbers. Depending on how your child was taught in elementary school, these calculations may be simple for him. You can always check on a calculator, or by using the old pencil-and-paper method.

Multiplying in Base-Ten

When faced with a problem like 120×8, many students instinctively reach for a calculator. But multiplying multi-digit numbers can be a breeze if you break numbers down into their base-ten components.

120 is the same as $100 + 20$. So 120×8 is the same as $100 \times 8 + 20 \times 8$, or $800 + 160$, which is 960. No calculator required!

Examples

1. 45×7
Break 45 down to its parts. $40 \times 7 = 280$, and $5 \times 7 = 35$, so we add $280 + 35 = 315$.

2. 639×4
$600 \times 4 + 30 \times 4 + 9 \times 4$
$2{,}400 + 120 + 36$
$639 \times 4 = 2{,}556$

If this seems like a lot to keep track of in your head at one time, it's probably because you haven't practiced it much yet. With time and patience, you and your child can become adept at these mental calculations. Don't think you can do it? You're capable of more than you know, and your child is, too.

3. 20×7
We know that $2 \times 7 = 14$. Since 20 is 10 times larger than 2, we can find 20×7 by simply making 14 ten times larger—that is, $20 \times 7 = 140$. Many students will explain this problem by saying that you just need to multiply 2 by 7 and "add" a zero. If you hear your child say that, be cautious about the language he is using—adding zero does not change a number, and technically we are not adding anything here.

In Chapter 5, we will learn a method for using geometry to multiply two multi-digit numbers together. While this might be a bit too complex for most people to do mentally, we can still apply the same ideas we have just seen to a problem like the next example.

4. 18×57

In this case, we need to break one of the two-digit numbers up first. It doesn't matter which, so let's go with 57.

$18 \times 57 = 18 \times 50 + 18 \times 7$

Maybe you can't do 18×50 or 18×7 in one step in your head (I can't either . . . haven't practiced that enough yet). But you can make it more manageable by now breaking the 18 into *its* parts:

$18 \times 50 = 10 \times 50 + 8 \times 50$, and
$18 \times 7 = 10 \times 7 + 8 \times 7$

Now we can put it all together:

$18 \times 57 = 10 \times 50 + 8 \times 50 + 10 \times 7 + 8 \times 7$
$18 \times 57 = 500 + 400 + 70 + 56$
$18 \times 57 = 1{,}026$

These are a lot of numbers to keep track of in your head at this point. Don't worry. With a little bit of practice, you can still compute it faster on paper than you can on a calculator.

Dividing in Base-Ten

Of the four basic operations, division is the most difficult to perform mentally. While many division problems can be easily solved by a student who is familiar with multiplication facts, your child will see many division problems throughout his math education that will require either paper and pencil or a calculator. It is worthwhile to develop the skill of recognizing when a problem can be done quickly in your head and when it cannot.

First and foremost, it is critical that your child have his multiplication facts memorized by now, at least up to 10×10. If he has not memorized them yet, all you can do is practice, practice, practice. There should be no need to peek at the multiplication chart in the back of his notebook. If your child has those basic facts memorized, then there are several types of division problems he should be able to perform mentally without too much trouble:

- Basic facts problems
- Extensions of basic facts
- Breaking up the dividend
- Using factors of the divisor

Let's take a quick look at each of these problem types.

Basic Facts Problems

Once your child has memorized all of her basic multiplication/division facts, problems like $24 \div 6$ or $32 \div 8$ should never require a calculator or even more than a second's thought.

Extensions of Basic Facts

If $24 \div 6$ is a simple recall problem, then $240 \div 6$ should be just as simple—it is almost identical, except for place value. If $24 \div 6 = 4$, then $240 \div 6 = 40$. 240 is 10 times as much as 24, so $240 \div 6$ is 10 times as much as $24 \div 6$. Similarly, $240 \div 60 = 4$.

Examples

1. $420 \div 7 = 60$ because $42 \div 7 = 6$

2. $14{,}400 \div 12 = 1{,}200$ because $144 \div 12 = 12$

3. $2.8 \div 4 = 0.7$
We have not dealt with decimals yet, but in this problem 2.8 is one tenth as large as 28. Since there are 7 fours in 28, there is $\frac{7}{10}$ of a 4 in 2.8.

Breaking Up the Dividend

Sometimes the dividend (the number being divided) can be broken up into parts, where the digits are multiples of the divisor. Let's see some examples.

Examples

1. $248 \div 2$

This problem is a lot simpler than it looks because each of the digits in 248 is divisible by 2. If we deconstruct the dividend (248), the division is easy:

$248 = 200 + 40 + 8$
$200 \div 2 = 100$, $40 \div 2 = 20$, and $8 \div 2 = 4$, so
$248 \div 2 = 100 + 20 + 4 = 124$

2. $128 \div 4$

In this case, 4 doesn't go into 1, but it does go into 12 and it goes into 8, too. So we can break the dividend up into two components that are multiples of 4:

$128 = 120 + 8$
$120 \div 4 = 30$, and $8 \div 4 = 2$, so
$128 \div 4 = 30 + 2 = 32$

3. $2{,}849 \div 7$

$2{,}849 = 2{,}800 + 49$.
$2{,}800 \div 7 = 400$, and $49 \div 7 = 7$,
so $2{,}849 \div 7 = 400 + 7 = 407$

The above examples have all relied upon our understanding of the base-ten system.

4. $98 \div 7$

In this case, we cannot simply break 98 into its place value components because 7 does not go into 90 or 8. But we can break it up in another way that will work well for a quick mental calculation:

$98 = 70 + 28$
$70 \div 7 = 10$ and $28 \div 7 = 4$, so
$98 \div 7 = 10 + 4$
$98 \div 7 = 14$

5. $152 \div 8$

In this case, let's think of multiples of 8 that end in a 2. The multiples of 8 are 8, 16, 24, **32**, 40, 48, 56, **72**, etc. Let's use 72, since it is nice and big:

$152 = 72 + 80$
$72 \div 8 = 9$
$80 \div 8 = 10$
So $152 \div 8 = 19$

Of course, you could also start thinking of this problem by taking out 80—since it is 10×8, it is a really good place to start. $152 - 80 = 72$, so again we will have 10 and 9 to add together.

6. $387 \div 9$
We know that $9 \times 4 = 36$, so let's break 387 into $360 + 27$.

$360 \div 9 = 40$
$27 \div 9 = 3$
So $387 \div 9 = 43$

Using Factors of the Divisor

Some numbers may not be so easy to divide, but using factors can be a big help. For example, it may not be easy to divide a large number by 6, but the factors of 6 are 2 and 3, so it might be easier to divide by 2 and 3. Dividing by 6 is the same as dividing by 2 and by 3.

Examples

1. $150 \div 6$ may not be immediately obvious, but you know that $6 = 2 \times 3$. Let's divide 150 by 3 and then by 2:

$150 \div 3 = 50$ (since $15 \div 3 = 5$)
$50 \div 2 = 25$ (easy if you think about money)
So $150 \div 6 = 25$

It works the same if you divide by 2 first:

$150 \div 2 = 75$
$75 \div 3 = 25$
So $150 \div 6 = 25$

2. $96 \div 6$ works in the same way:

$96 \div 3 = 32$
$32 \div 2 = 16$
So $96 \div 6 = 16$

If $32 \div 2$ is not obvious to you, try using one of the tricks we have already covered.

3. $375 \div 15$
We know that $15 = 3 \times 5$. So if we can divide 375 by 3 and 5, we can solve it. $375 \div 3$ can be done by breaking up the dividend, since $300 \div 3 = 100$ and $75 \div 3 = 25$. So $375 \div 3 = 125$. Now we divide 125 by 5, which is also very simple: $125 \div 5 = 25$.

$375 \div 3 = 125$
$125 \div 5 = 25$
So $375 \div 15 = 25$

Long Division

Sometimes it just isn't reasonable or efficient to try to divide in your head. When this happens, the calculator may be a good friend. But for your child, who is still developing her basic number sense, a better strategy is to stick with pencil-and-paper. By practicing the long division algorithm, your child will strengthen her understanding of the number system and notice some things about numbers that will be critical for her future math work.

How Does Long Division Work?

First, let's divide a simple three-digit number by a one-digit number. We will use the idea of place value to make sense of how and why the algorithm works. This example is one of the simpler types of problems to solve mentally, since each digit of the divisor is divisible by the dividend. That simplicity will help us understand how the algorithm works.

$$864 \div 2$$

Before we use the long division algorithm, let's think about the number 864. 864 can be rewritten as $800 + 60 + 4$. If we can divide each of these parts by 2, then we can add those pieces back together and find the quotient.

$$800 \div 2 = 400$$

$$60 \div 2 = 30$$

$$4 \div 2 = 2$$

So in total, $864 \div 2 = 400 + 30 + 2 = 432$.

Let's see how the same ideas appear in the long division algorithm. 864 is the dividend, or the number being divided, which goes inside the "division house." 2 is the divisor, or the number doing the dividing, which goes outside.

$$2\overline{)864}$$

The first step is to find out how many times 2 goes into 8. Since 2 × 4 = 8, we write 4 on top of the house, and subtract 8 from the 8 in in the dividend. That leaves nothing (i.e., zero), so we bring down the 64.

```
    432
 2 )864
   -8↓↓
   ----
     64
```

What have we really done here? The language in the previous paragraph is very informal. Many teachers use this kind of language, but it does not precisely describe what is being done. When we find that 2 goes into 8 four times, what we are really asking is how many times 2 goes into *800*, and the 4 that we write on top of the division house really means *400*.

Similarly, when we take the next step, we are thinking about how many 2s are in 60, not 6. Here is the problem, worked out completely:

```
    432
 2 )864
   -8↓
   ---
     6
    -6↓
    ---
      4
     -4
     --
      0
```

To summarize, what we are doing when we use the long division algorithm is taking it in steps. First, we find out how many times the divisor goes into the largest place value of the dividend. Then, we see what's left and go from there. We found out that 2 goes 400 times into 800. But 2 goes into 864 more than 400 times, so we subtract 2 × 400 from the dividend to see how much is left to divide.

We still need to figure out how many times 2 goes into the remaining 64. First, 2 goes 30 times into 60, so we subtract 2 × 30 from 64, leaving 4. Then, 2 goes 2 more times into the remaining 4, so the total number of 2s in 864 is 432.

Here's another example:

$$192 \div 6$$

In this case, 6 doesn't go into 1, so we look at the 1 and the 9 together as 19. That's an informal way of saying that we will decompose the number (break up the dividend) this way:

$$192 = 190 + 2$$

This is one of our mental math strategies from earlier, and it is the basis of the long division algorithm. Here's the long division algorithm:

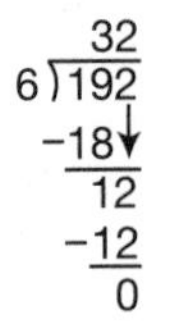

When we subtract the 18 from 19, the "1" that is left over is really 10, because we have really subtracted 180 from 190. That leaves 10, which we add to the 2 ones in the dividend to make 12. $6 \times 2 = 12$, so we write 2 as the ones digit in the quotient.

What happens when the divisor is more than one digit long?

$$1{,}092 \div 26$$

In this case, when we set up the long division algorithm, we may need to use three digits before we can easily find a multiple of 26. Since 26 doesn't go into 1 or 10, we must try 109. We know that $25 \times 4 = 100$, so 26×4 must be pretty close to 109—in fact, it's 104. Here's the algorithm worked out:

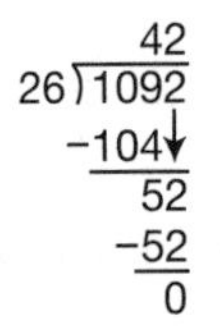

What we have done, essentially, is break down the dividend again:

$$1{,}092 = 1{,}040 + 52$$

$$40 \times 26 = 1{,}040$$

$$2 \times 26 = 52$$

$$\text{So } 1{,}092 \div 26 = 42$$

Summary So Far

So far in this chapter, we have used the base-ten system as a tool for efficiently performing the four operations on whole numbers. By connecting the concept of place value to the algorithms, we have attempted to make sense of these age-old methods, which many adults are accustomed to performing without understanding. Now we will extend these ideas to decimals, which are just another part of the base-ten system. Along with fractions, decimals make up a significant majority of the "number sense" curriculum in upper elementary and middle schools.

Decimals

In Chapter 2, we did a lot of work with fractions, which are numbers in between whole numbers. The base-ten system gives us another way to represent these numbers by using decimals to extend our base-ten system to smaller and smaller place values. You probably already know from dealing with money that .5 is a half and that .25 is a fourth or a quarter. But many students never really grasp the meaning of decimals as an extension of the base-ten system. With whole numbers, every digit has 10 times as much place value as the digit to its right. It works in the opposite way as well—every time you move one digit to the right, the place value *divides* by 10. So 200 is one-tenth of 2,000, and 20 is one-tenth of 200. 2 is one-tenth of 20. 0.2 is one-tenth of 2, and 0.02 is one-tenth of 0.2.

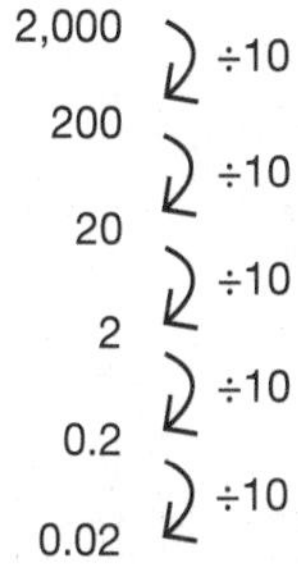

2,000
 ↓ ÷10
200
 ↓ ÷10
20
 ↓ ÷10
2
 ↓ ÷10
0.2
 ↓ ÷10
0.02

Refer to the following chart to help remember some of the most common place values. No matter how many digits a number has, the place values never change. The first digit to the left of the decimal point is always the ones, and the one to the left of that is the tens, etc. The first digit to the right of the decimal is always tenths, followed by hundredths, etc. If a number is written without a decimal point, then it is a whole number, and the final digit is the ones digit.

9	0	3,	4	3	1,	5	9	2	.	9	5	0	7	4	1	6
Hundred Millions	Ten Millions	Millions	Hundred Thousands	Ten Thousands	Thousands	Hundreds	Tens	Ones		Tenths	Hundredths	Thousandths	Ten Thousandths	Hundred Thousandths	Millionths	Ten Millionths

Converting Fractions to Decimals

Since decimals and fractions both represent numbers between whole numbers, they can be used in many of the same ways—every fraction has an equivalent decimal form. The long division algorithm can be used to convert any fraction to its decimal form, which is a great way to practice the long division algorithm—and to understand its value. Recall from Chapter 2 that a fraction is a quotient. For example, $\frac{1}{4}$ is the answer to 1 ÷ 4. The numerator is the dividend, which goes inside the division box, and the denominator is the divisor, which goes outside. In the case of $\frac{1}{4}$, this may look uncomfortable to your child at first, since he's used to seeing the larger number inside the house. But $\frac{1}{4}$ is a fraction, and it's smaller than 1, so when we divide it, we are breaking a whole into several pieces.

$$4\overline{)1}$$

Of course, 4 doesn't go into 1 at all, so we have to think of 1 in a different way. The number 1 is not just 1 one. It is also 10 tenths, or 100 hundredths. We can write any whole number with as many zeros as we want after the decimal point because a whole number has nothing in these smaller place values. If you aren't sure how many 0s to use, it's always safe to just write a bunch. You can always write more later if you need them, and if you ever get a remainder of 0 in your long division algorithm, then you're done, no matter how many 0s follow.

$$4\overline{)1.0000}$$

4 doesn't go into 1, but 4 goes into 10 two times. But using our knowledge of place value, that isn't really 10—it's 10 *tenths*. So 4 doesn't really go in 2 times, it goes in 2 *tenths* of a time. So we write .2 above:

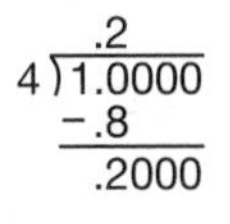

Then we multiply and subtract to find out what's left. 2 tenths × 4 = 8 tenths:

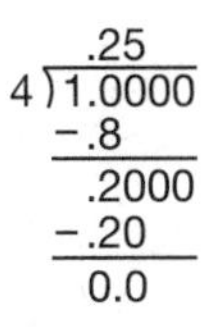

4 doesn't go into 2, but it goes into 20 five times. But that's not really 20, it's 20 *hundredths*, and 4 goes into 20 *hundredths* 5 *hundredths* of a time. So we write 5 hundredths on top, multiply, and subtract:

$$
\begin{array}{r}
.25 \\
4\overline{)1.0000} \\
-.8 \\
\hline
.2000 \\
-.20 \\
\hline
0.0
\end{array}
$$

Now there's no remainder, and we have our decimal equivalent: $\frac{1}{4}$ = .25.

When dealing with tenths and hundredths, it is simple to represent a problem visually, which helps students understand that decimals and fractions really represent the same thing. In the following diagram, the square has been divided into 100 equal-sized pieces. The thicker lines show how the square can be divided into 4 equal pieces, one of which is shaded. There are 25 hundredths inside the shaded area.

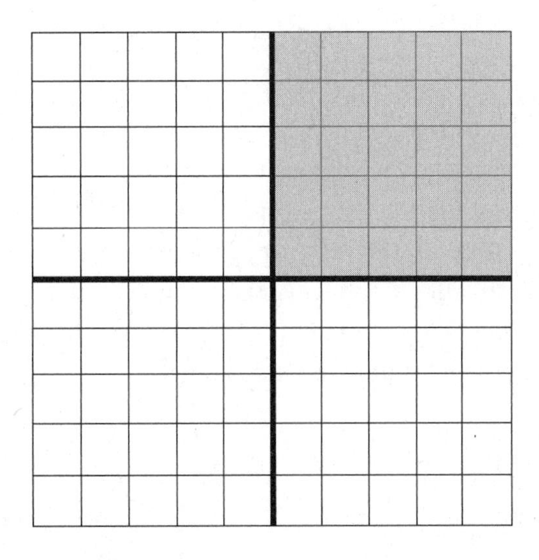

Conveniently, we can use this equivalent to find the decimal forms of $\frac{2}{4}$, $\frac{3}{4}$, and $\frac{4}{4}$. Since $\frac{2}{4}$ is $2 \times \frac{1}{4}$, $\frac{2}{4} = 2 \times .25 = .50$. In the same way, $\frac{3}{4} = .75$ and $\frac{4}{4} = 1.00$.

Repeating Decimals

Sometimes the fraction-to-decimal conversion does not work out so cleanly. When a decimal ends after a certain number of digits, like .25 or .1039, we call it a *terminating decimal.* But some fractions, when converted, never end. They create a repeating pattern of digits, known as a *repeating decimal.*

Examples

1. Perhaps the most well-known repeating decimal is the decimal form of $\frac{1}{3}$.

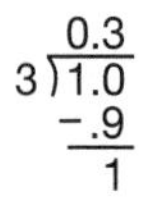

At this point, we have thought of 1 as 10 tenths, and found that subtracting 0.3×3 from the whole leaves a remainder of one tenth (that 1 at the very bottom). Now we express the 1 tenth as 10 hundredths (informally, we are "adding a zero"), and keep the process going:

```
   0.33
3)1.00
  -9
   10
   -9
    1
```

At the end of this step, we are again left with a 1—this time 1 hundredth, which we can express as 10 thousandths, which will give us another 3 in the quotient, and another 9 to subtract from another 10, leaving another 1. At this point, we should see the pattern will go on forever without ever changing. So $\frac{1}{3} = 0.3333333333333333333 \ldots = 0.\overline{3}$. We use the bar above the 3 to show that that digit repeats forever.

Not every repeating decimal repeats from the beginning, and some may have more than one digit repeating, as in the following examples:

2. $\frac{7}{12}$

$$\frac{7}{12} = 12\overline{)7.000} \quad \begin{array}{r} 0.583 \\ 12\overline{)7.000} \\ -6\,0 \\ \hline 1\,00 \\ -96 \\ \hline 40 \\ -36 \\ \hline 4 \end{array}$$

Now that the remainder of 4 has repeated, the 3 in the quotient will also repeat: $\frac{7}{12} = 0.583333\ldots = 0.58\overline{3}$.

3. $\frac{8}{11}$

$$\frac{8}{11} = \begin{array}{r} 0.72 \\ 11\overline{)8.00} \\ -7\,7 \\ \hline 30 \\ -22 \\ \hline 8 \end{array}$$

There's the 8 again—this decimal will repeat two digits: $\frac{8}{11} = 0.727272727272\ldots = 0.\overline{72}$.

BENCHMARK DECIMALS

In middle school, your child will need to convert fractions to decimals very often. It helps to have a few common ones memorized so that you can easily generate some others. Here are a few of the most important unit fractions:

$\frac{1}{2} = 0.5$	$\frac{1}{5} = 0.2$
$\frac{1}{3} = 0.\overline{3}$	$\frac{1}{10} = 0.1$
$\frac{1}{4} = 0.25$	$\frac{1}{100} = 0.01$

You can use the fractions above to find the decimal equivalents of many other fractions that have the same denominator.

Examples

1. $\frac{2}{3}$ is twice as much as $\frac{1}{3}$, so

   $\frac{2}{3} = 2 \times \frac{1}{3}$

   $\frac{2}{3} = 2 \times .333333\overline{3}$

   $\frac{2}{3} = .666666\overline{6}$

2. $\frac{7}{5} = 7 \times \frac{1}{5} = 7 \times .2 = 1.4$, so $\frac{7}{5} = 1.4$. Since the numerator is larger than the denominator, the fraction's value is greater than 1.

3. $\frac{29}{100}$ is $29 \times .01$, which is .29.

Operating with Decimals

Now that we understand what decimals represent and where they come from, let's look at how to operate with decimals. Fortunately, since decimals are simply another part of our base-ten system, operating with decimals works exactly the same way as operating with whole numbers. The important thing is to always pay attention to place value.

Adding Decimals

When adding two decimals, simply add each place value, just as you would for whole numbers. You can do this with a variety of methods—the traditional algorithm, decomposing the numbers, or with visual representations.

Examples

1. $9.61 + 4.38$
   Here it is, decomposed:

		ones		tenths		hundredths		
9.61	=	9	+	0.6	+	0.01		
+ 4.38	=	4	+	0.3	+	0.08		
		13	+	0.9	+	0.09	=	13.99

Your child may be familiar with place value representations where a large square is used to represent one whole, a thin strip represents one tenth, and a small square represents one hundredth, as shown below. These are related to some of the fraction representations we used in Chapter 2.

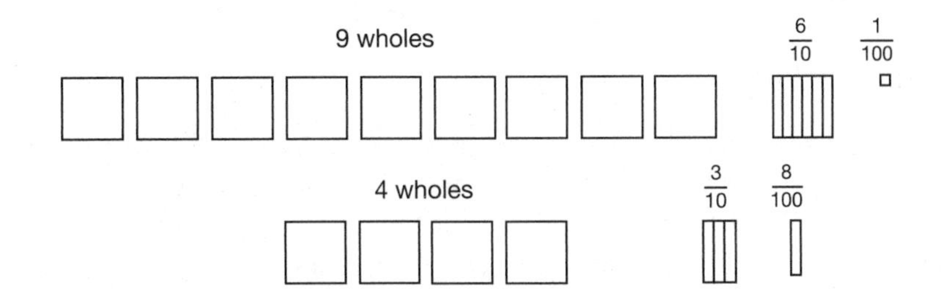

Altogether, we have 13 wholes, $\frac{9}{10}$, and $\frac{9}{100}$, or 13.99. The standard algorithm works just as well here:

$$\begin{array}{r} 9.61 \\ +\,4.38 \\ \hline 13.99 \end{array}$$

2. 12.75 + 3.88

In this problem, we will end up with more than 10 in each place value. When we decompose the numbers, we will have to rearrange things a little bit. Again, these are all steps that could be done mentally, with a little practice.

$$\begin{array}{rcl} 12.75 & = & 10 + 2 + 0.7 + 0.05 \\ +\ 3.88 & & + 3 + 0.8 + 0.08 \\ \hline & & 10 + 5 + 1.5 + 0.13 \\ & & 10 + 5 + 1 + 0.5 + 0.1 + 0.03 \\ & & 10 + 6 + 0.6 + 0.03 \\ & & \boxed{16.63} \end{array}$$

Using the traditional algorithm, the rearranging that we did—converting 1.5 to 1 + 0.5 and 0.13 to 0.1 + 0.03—is done by "carrying the 1":

$$\begin{array}{r} {}^{1\,1} \\ 12.75 \\ +\ \ 3.88 \\ \hline 16.63 \end{array}$$

Neither method is considered better than the other, but many students do find it difficult to keep track of carrying the 1 in their heads.

3. 4.82 + 12.765

One very common error students make when trying to use the traditional algorithm occurs when adding two decimals that have different numbers of digits after the decimal point, as in this example. The important fact to remember is that you must add tenths to tenths, hundredths to hundredths, etc. So when setting up the addition algorithm, line the decimals up vertically. Since 4.82 has no thousandths place, you can write a 0 there to make it easier to show the process—82 hundredths is the same as 820 thousandths: $\frac{82}{100} = \frac{820}{1{,}000}$.

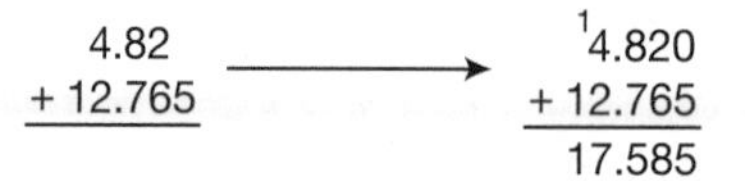

By decomposing the numbers, you can see that 4.82 has no thousandths to add:

$$\begin{array}{r} 4.82 \\ +\ 12.765 \\ \hline \end{array} = \begin{array}{rcrcrcrcr} & & 4 & + & 0.8 & + & 0.02 & & \\ +\ 10 & + & 2 & + & 0.7 & + & 0.06 & + & 0.005 \\ \hline 10 & + & 6 & + & 1.5 & + & 0.08 & + & 0.005 \end{array}$$

$$\boxed{17.585}$$

Subtracting Decimals

The way place value is used in addition works just as well for subtraction because they are inverse operations. Borrowing works in the same way for decimals as it does for whole numbers. Again, the important thing is to make sure the decimal points are lined up so that all the place values are together.

Example

Let's subtract 4.83 – 1.732. Since 4.83 has no thousandths place, we can put a 0 in that place and line up our decimal points.

$$\begin{array}{r} {\scriptstyle 1} \\ {\scriptstyle 7\ 2\ 1} \\ 4.\not{8}\not{3}\not{0} \\ -\ 1.732 \\ \hline 3.098 \end{array}$$

This is where the expanded form can be really helpful. Let's break the numbers up, but not all the way:

$$\begin{array}{r} 4.83 \\ -\ 1.732 \\ \hline \end{array} = \begin{array}{rcrcr} 4 & + & 0.8 & + & 0.030 \\ -\ 1 & - & 0.7 & - & 0.032 \\ \hline \end{array}$$

So we are subtracting 1 whole from 4 wholes, 7 tenths from 8 tenths, and 32 thousandths from 30 thousandths. Now we can convert 8 tenths and 30 thousandths to 7 tenths and 130 thousandths. In other words, we are taking 1 tenth from the 8 tenths, and turning it into 100 thousandths:

$$\begin{array}{r} 4.83 \\ -\ 1.732 \\ \hline \end{array} = \begin{array}{rcrcr} 4 & + & 0.8 & + & 0.030 \\ -1 & - & 0.7 & - & 0.032 \\ \hline 4 & + & 0.7 & + & 0.130 \\ -1 & - & 0.7 & - & 0.032 \\ \hline 3 & + & 0.0 & + & 0.098 \end{array}$$

$$\boxed{3.098}$$

Multiplying Decimals

We have seen that basic multiplication facts can be used to multiply with multiples of 10. For example, 50 × 9 = 450 because 50 is 10 times as much as 5, and 5 × 9 = 45. Similarly, 60 × 30 = 1,800. You may remember a rule about "adding the zeros," but we are not really adding anything. 60 is 6 × 10, and 30 is 3 × 10, so 60 × 30 = 6 × 10 × 3 × 10, which is the same as 18 × 100.

The same strategy works with decimals.

Examples

1. 0.4 × 8

0.4 is 4 tenths. We know that 4 × 8 = 32, so .4 × 8 is 32 *tenths*, which is 3 wholes and 2 tenths.

0.4 × 8 = 3.2

We could show this with a simple number line, which your child has likely seen in elementary school:

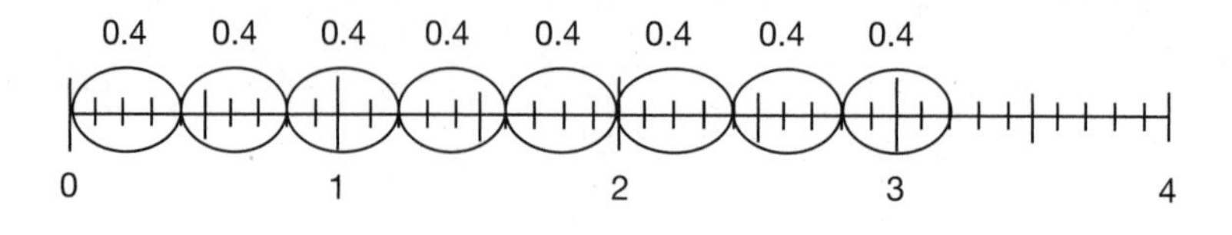

2. 0.9×0.7

There is a lot to learn from this simple-looking problem. We will solve it in a few different ways because multiple methods will help you and your child continue to build strong number sense and fluency.

First, let's solve it as a fraction multiplication problem. We know that these decimals can be converted to fractions.

$0.9 \times 0.7 = \frac{9}{10} \times \frac{7}{10}$

After working through Chapter 2, you know how to multiply these:

$\frac{9}{10} \times \frac{7}{10} = \frac{63}{100}$

We can turn this answer back into decimal form. $\frac{63}{100} = 0.63$, so $0.9 \times 0.7 = 0.63$.

Another way to solve it is by using an area model, as we did in Chapter 2. Here is a whole, cut into tenths, both vertically and horizontally, making a total of 100 equal-sized pieces. $\frac{7}{10}$ are shaded horizontally, and $\frac{9}{10}$ are shaded vertically.

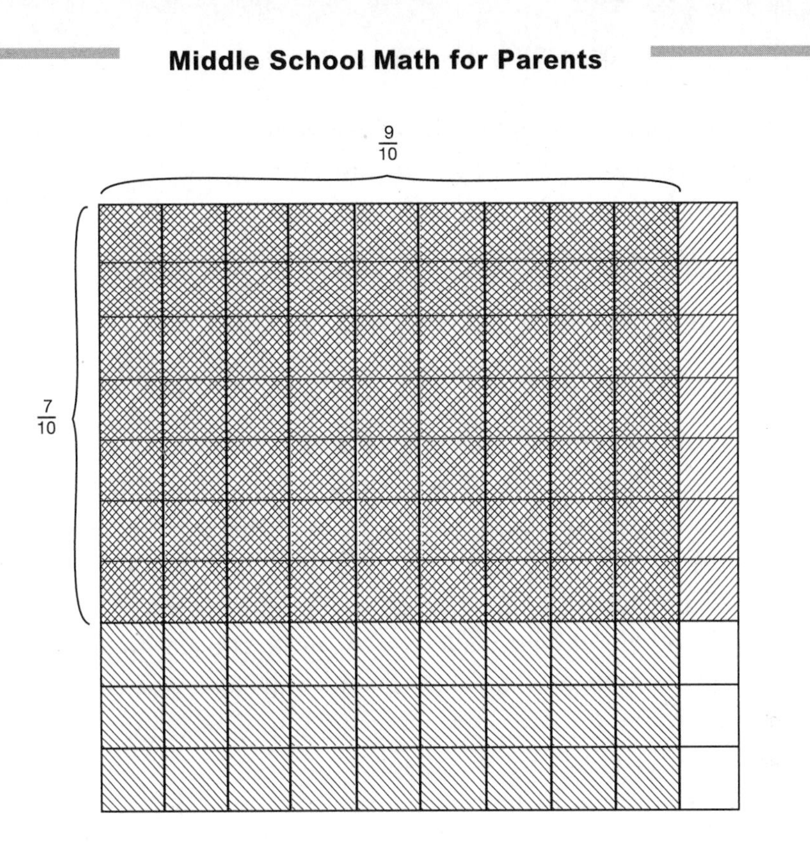

Altogether, $\frac{63}{100}$ are double-shaded, so 0.9 × 0.7 = 0.63.

A third method is to again use place value and basic multiplication facts. 0.9 is the same as 9 ÷ 10, and 0.7 is 7 ÷ 10. We can multiply the 9 and the 7 together to get 63. Since each of these factors was originally being divided by 10, we can divide the 63 by 10 twice. 63 ÷ 10 = 6.3, and 6.3 ÷ 10 = 0.63.

Finally, we have the traditional algorithm:

$$\begin{array}{r} 0.9 \\ \times\ 0.7 \\ \hline 0.63 \end{array}$$

The little squiggles under the digits indicate the old rule: Count the number of digits after the decimals in the factors, and put that many digits after the decimal in the product. It's a simple rule, but like most rules in math, it is hard to remember and easy to mess up. What if the last digit is a 0, for instance? Does that count?

It's better to use a little bit of number sense to decide where the decimal point should go. Start by multiplying the numbers, and write the product with no decimal point:

$$\begin{array}{r} 0.9 \\ \times\, 0.7 \\ \hline 63 \end{array}$$

Now we know that the product will have a 6 and a 3, but we don't know the exact place values of the digits: it could be 63 or 6.3 or 0.63 or 0.063, etc. We do know that we are multiplying two numbers together that are both smaller than 1. And when you multiply by a fraction or decimal smaller than 1, you make the other number smaller. So the product needs to be smaller than both 0.9 and 0.7. 0.63 fits the bill. 0.063 would be too small.

Let's try this method again on another example:

3. Consider 5.82 × 97.5. Let's first ignore the decimal points for a moment and solve the problem as if the numbers were both whole numbers, 582 × 975:

$$\begin{array}{r} {\scriptstyle 7\ 1} \\ {\scriptstyle \not{3}\ \not{1}} \\ {\scriptstyle \not{4}\ \not{1}} \\ 582 \\ \times\ 975 \\ \hline {}^{2}2{,}910 \\ 40{,}740 \\ 523{,}800 \\ \hline 567{,}450 \end{array}$$

So 582 × 975 = 567,450.

CONNECTING THE METHODS

This algorithm is just another way of using decomposed numbers. In order to multiply 582 by 975, we broke 975 into 900 + 70 + 5.

$5 \times 582 = 2{,}910$

$70 \times 582 = 40{,}740$

$900 \times 582 = 523{,}800$

By adding these three products together, we get the total. For many parents, the zeros that are placed in this algorithm when multiplying by the 7 and the 9 are a mystery. But they make sense if you think about place value—the 9 is really 900, and the 7 is really 70.

So we've found the product of 582 and 975. But the problem asked us to multiply 5.82 and 97.5. The digits 567450 will be in the product. If we round both of the factors to something convenient, we can figure it out easily. 5.82 is pretty close to 6, and 97.5 is close to 100. Multiplying 6×100 is simple—600. So the product of 5.82×97.5 should be *pretty close* to 600. This makes it easy to find the right place for the decimal point. $5.82 \times 97.5 = 567.45$.

Dividing Decimals

Rounding to find the correct place for the decimal point works for division as well.

Examples

1. $7.68 \div 1.2$

```
     64
1.2)7.68
   -7 2↓
     48
```

(Note that all the numbers underneath the dividend of 7.68 completely ignore the decimal point and place value—this is a shorthand method, and placing the decimal points properly can often make this representation more confusing for students).

Now we must place the decimal point. Is it 64? 6.4? 0.64? Let's think about the numbers. 7.68 is close to 7 or 8. 1.2 is close to 1. How many times does 1 go into 7 or 8? It's certainly much closer to 6 than to 60 or 0.6. Therefore, our quotient is 6.4.

It is always possible that division may result in a repeating decimal. The strategy remains the same.

2. $72.1 \div 4.2$

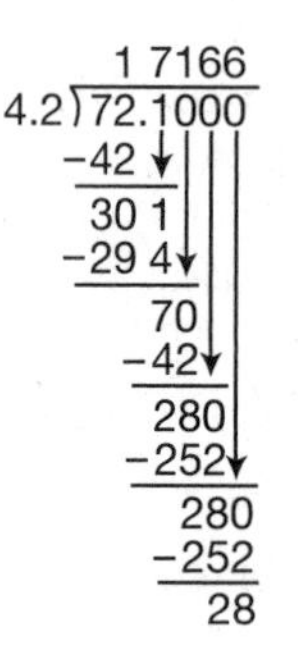

Now that the 28 is repeating in the remainder, the 6 in the quotient will repeat forever. So where does the decimal point go? 4.2 is close to 4, and 72 is close to 80. $4 \times 20 = 80$. Is the quotient $1.716\overline{6}$? $17.166\overline{6}$? $171.66\overline{6}$? $17.166\overline{6}$ is closest, so we place the decimal point after the 7.

Summary

The four operations—addition, subtraction, multiplication, and division—are the basis of all mathematics. There are many ways to perform these operations, and a good mathematician has the skill and flexibility to choose the best method for a given situation and use it. By understanding the base-ten number system and the power of place value, your child will be empowered to try a variety of methods. Visual representations will also help with this fluency, as we have seen several times in this chapter. Of course, there will be times when the old paper-and-pencil methods will be required, and that's why it is important for your child to get in some good practice on the traditional algorithms for all four operations—particularly division, which is the most challenging.

The most important thing you can do as a parent is to help your child make good decisions about when and when not to use a calculator. While he is still developing proficiency with the four operations, it is definitely *not* the time to use a calculator—it will hold your child back from getting the knack of the operations. Once he has mastered the operations, and he is using them as a tool for more advanced mathematics, then the calculator can do no harm.

Throughout the book so far, we have used a variety of approaches to each topic, many of which were likely unfamiliar to you before. Many parents find themselves confused as to why all these different methods are necessary. In Chapter 4, we will address that concern as we study several different ways to solve proportions, one of the most important early algebra concepts.

Learning the "New Way"

Why Ratios and Proportions Make More Sense Now Than Ever

Math Hasn't Changed

Parents sometimes wonder why schools need to purchase updated math textbooks so often. They don't have a problem with new history textbooks, because a lot can change in a few years; what good is a U.S. history textbook that doesn't include the current president? But when it comes to math, what has changed? 1 + 1 is still 2, and the area of a circle is still πr^2. While mathematical research is ongoing, the new discoveries are seldom, if ever, relevant to middle school or high school curricula. So other than the usual wear-and-tear, why do math textbooks need to be replaced?

It isn't the math that has changed—it's everything else. The incredible growth of technology since the turn of the millennium has changed the way we use and think about math. Some studies have predicted that large percentages of careers that today's middle-school students will be pursuing have not even been *invented* yet—that's how fast technology is changing our world. Even jobs that never had anything to do with math before are now heavily computer-driven. Car mechanics connect the vehicles they work on to a computer. Carpenters, electricians, and plumbers show up to jobs with tablets or laptops. Algebra is the language of computers, so for today's students, success in high school math courses is a basic requirement—not just for graduation, but for nearly any career that will interest them. Without algebra, our students will have very limited options later in life.

With that evolution, we have had to change the way we teach math. It used to be okay if only the top students took high-level math courses. Now it is imperative to get every student to reach those levels. And the level of challenge in those courses has not been reduced to accommodate everybody—if anything, the rigor has been raised as standards are periodically revised, pushing content to lower and lower grades, as the Common Core has done with many topics. Where Algebra 1 used to be the standard 9th-grade course, it is now an expectation for many 8th graders. Students are learning about fractions as early as kindergarten or 1st grade, rather than in upper elementary school.

As students are now being held to higher standards, it has been crucial for teachers to adapt their instructional methods. As a parent, you may remember math classrooms with desks in neat rows, where students copied the teacher's examples off the board, then silently solved long series of practice problems. Now your child may be in a math classroom that is rarely silent, where students are seated in groups and encouraged to tackle problems together, with little guidance from the teacher. In many classrooms, the lesson rarely comes to the point where the teacher presents a solution—it all comes from the students. And, rather than learning one way of solving a specific type of problem, students often come out of a lesson with several different ways of approaching a concept.

To many parents, this seems like a backwards, inefficient way of teaching. Parents often ask the question, "Why can't they just teach it the old way?" If this question has crossed your mind, I would encourage you to think about yourself and your high school classmates. How many of the people in your 10th grade math class have grown up to consider themselves "math people"? How many adults do you know who feel comfortable using fractions in daily life, or who break into a cold sweat at the thought of helping their children with algebra homework? The old teaching methods worked very well for a small number of students. They are insufficient for today's students and the needs of their generation.

Following a Recipe vs. a Chef's Craft

The "old way" was a *procedural approach*. That is, students were taught a series of procedures to follow in order to solve specific problems. Think of it like a cookbook, where there is a different recipe for every dish you might want to cook. Students may become very good at following specific steps to solve equations like $4x + 7 = 31$. But then if you throw something like $19 - 2x = 7$ at them, they either get stuck and shut down or they enter blindly into a familiar procedure and make a mess of the algebra because they are applying the wrong procedure at the wrong time.

If you go into a good restaurant, the chef is not using a cookbook night after night to come up with the specials and new dishes on the menu. The chef is *creating* those recipes from her knowledge of flavors, ingredients, and techniques, using some creativity and imagination.

A good mathematician is like a chef. She doesn't constantly scratch her head and ask, "Which type of problem is this? How do I solve it?" because nobody can realistically be expected to memorize the full cookbook of different math problems. Instead, a good mathematician uses a *conceptual approach*. Looking at the equation $19 - 2x = 7$, a good mathematician may ask questions like "What does this equation mean? If I'm subtracting $2x$ from 19 to get 7, how can I find out what x is? How can I rewrite this equation to get x by itself?" A good mathematician will often use her number sense to answer these questions. If $19 - 2x = 7$, then $19 - 7 = 2x$, so $2x$ must be 12, and $x = 6$. In today's math classrooms, students don't need their teachers to spend a day covering equations where the x term is subtracted because the teacher has already taught students how to make sense of equations, and the students have a whole collection of tools to use to interpret that equation and isolate x.

Proportions (and the Problem with a Procedural Approach)

"The product of the means is equal to the product of the extremes." Do you remember this forced rhyme intended to help students remember how to solve proportions? What kind of sense did that make when nobody knew what the words *means* and *extremes* meant in this context?

Take a moment to solve for x in the equation below. No spoilers—don't look at the explanation below until you've solved it.

$$\frac{6}{x} = \frac{14}{35}$$

If you remember learning this the same way most adults did, then you probably used cross-multiplication:

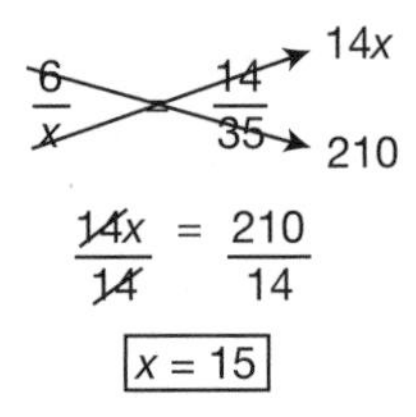

This method has been beaten into the heads of innocent students for countless generations. Asked to explain why it works, children and adults alike usually offer one of three responses: the blank stare, the nonsensical ramble, or worst of all, "Because my teacher said so." What's worse, many *math teachers* cannot explain why this method works!

But that's not the problem with the cross-multiplication method. The problem is the confusion it causes for students when approaching *other* types of problems.

Remember Chapter 2, when we learned how to multiply fractions?

$$\frac{3}{5} \times \frac{5}{6} = \frac{15}{30} = \frac{1}{2}$$

The numerators of the factors multiply together, and the denominators multiply together, then reduce if possible.

So now we get to the heart of the matter. When solving proportions, you *cross* multiply, but when multiplying fractions, you multiply *across*. Can you see how students would get confused? What ends up happening is a lot (I mean *a lot*) of this:

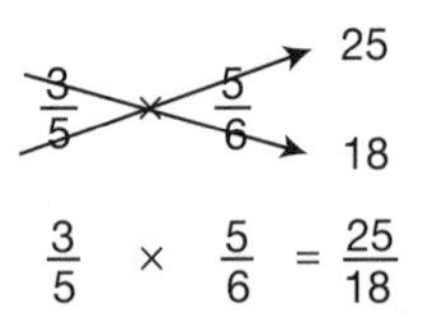

A student who has been taught conceptually can see right away that this answer is incorrect. When you multiply any number by a fraction smaller than 1, the result will be smaller than the original number. But here we have multiplied two fractions, both smaller than 1, and gotten an answer that is larger than 1. Impossible!

What's happening is that the student is *overgeneralizing* a concept she has learned. She has cross-multiplied so many times that as soon as she saw two fractions, she automatically used that method. If only she understood *why* cross-multiplication works, she would likely have been able to avoid that mistake.

In this chapter we will take the conceptual approach to ratios and proportions. We will study a variety of methods that will develop your child's proportional reasoning skills by building on number sense and pattern recognition, rather than a rote algorithm, so that you and your child will not need to depend on cross-multiplication, that old, incoherent and risky method.

Understanding Ratios

A *ratio* is a way of using a fraction to compare the size of two numbers. We use ratios every day, often without even being aware of it. When you put gas in your car or check your speedometer; when you mix ingredients for a recipe or estimate how much food you will need for a party; when you compare prices for different size packages of a favorite snack or decide how early to leave for an important appointment; when you check the news to see whether it is going to rain—these are all times when you use ratios.

The Language of Ratios

Simply stated, a ratio is a fraction. What makes ratios special is the way mathematicians use them to describe and compare things in a real-world context.

Example

One example that is always popular with middle school students is to compare the number of boys and girls in a classroom. Suppose there are 25 students in a class: 10 boys and 15 girls. There are several different ways of describing the ratios in this class.

The ratio of boys to girls is 10 to 15. This can also be written as 10:15 or $\frac{10}{15}$, although for reasons we will see in a moment, the fraction form does not make much sense here.

Some other ratios in this situation include:

- The ratio of girls to total students is 15 to 25.
- The ratio of boys to total students is 10:25.
- Boys constitute $\frac{10}{25}$ of the class.

Just like with regular fractions, it is often convenient to express a ratio in simplest form. If the ratio of boys to girls is 10 to 15, that can be reduced to an equivalent ratio of 2 to 3. In other words, for every 2 boys in the class, there are 3 girls (students tend to read more into that statement than it is meant to imply!). The class could be divided into groups of 5, with 2 boys and 3 girls in each group.

Two Types of Ratios

In the examples above, each ratio was made up of two of the three quantities in the problem: boys, girls, and total students. We can think of each quantity as either a *part* or the *whole*. Boys are a part, and Girls are a part, but Total Students is the whole.

There are two different types of ratios that are often useful for comparing things.

1. **Part-to-whole ratios** compare the size of one part to the size of the whole, just like a typical fraction where the numerator is the part and the denominator is the whole. The ratio of boys to total students and the ratio of girls to total students are examples of part-to-whole ratios.

2. **Part-to-part ratios** compare the sizes of two different parts of the whole. In this case, it usually doesn't make sense to use the fraction form, since the denominator does not represent the whole of anything. The ratio of boys to girls and the ratio of girls to boys are examples of part-to-part ratios.

Now we are ready to solve some problems. In our "new way" approach to this topic, we will look at four different methods for solving ratio and proportion problems.

Method #1: Ratio Tables

A ratio table is a handy tool for solving ratio problems and for helping students understand the patterns that emerge when ratios are used to solve problems. If your child masters ratio tables, he or she will be well prepared for much of the work to come in high school algebra.

Example

A classic example is a drink mix. You can think of any adult drinks you like, but we'll keep it family-friendly. To make her famous fruit punch, Jasmine mixes 5 cups of grape juice with 2 cups of cranberry juice. Jasmine can use the following table to figure out the ratios when she is mixing juice for large parties:

Grape Juice	5 cups	10 cups	15 cups	20 cups	25 cups
Cranberry Juice	2 cups	4 cups	6 cups	8 cups	10 cups

This table is helpful for finding equivalent ratios for larger quantities. A ratio of 5 cups to 2 cups is equivalent to a ratio of 15 cups to 6 cups, or 20 cups to 8 cups. All of these are part-to-part ratios. If Jasmine wants to know the total quantity of fruit punch she will be making, she can add another row to her table:

Grape Juice	5 cups	10 cups	15 cups	20 cups	25 cups
Cranberry Juice	2 cups	4 cups	6 cups	8 cups	10 cups
Total Fruit Punch	7 cups	14 cups	21 cups	28 cups	35 cups

Notice that the relationships between the columns in a ratio table are *multiplicative*. That means to get the numbers in one column from the numbers in any other column, you multiply both numbers in the original column by the same factor:

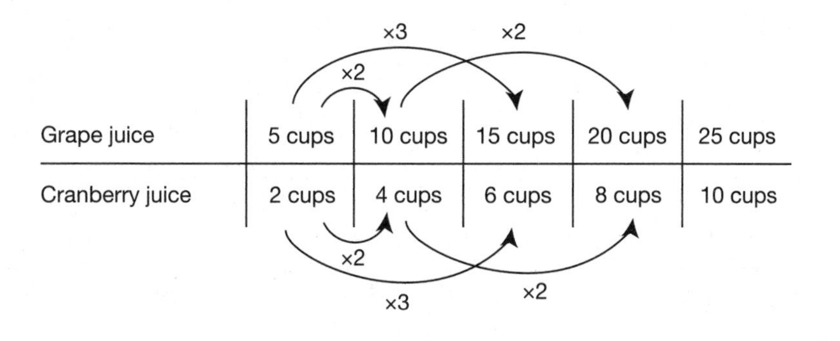

Many students have the misconception that this relationship should be additive—in other words, that you should add the same number to go from one column to the next. This works only if the intervals from one column to the next are all the same. Sometimes it is more convenient in a table to change the intervals. In the example below, tickets are being sold for a school concert. The table shows the total income for several numbers of tickets.

# of Tickets Sold	1	2	10	50	100	1,000
Total Income	$6	$12	$60	$300	$600	$6,000

In this example, we can still multiply one column by a certain number to get the numbers in another column—for example, 2 × 25 = 50, and $12 × 25 = $300—but there is no consistent interval from one column to the next.

Finding Missing Values

You can make a ratio table out of any ratio. Suppose you are buying pizzas for a birthday party. You have a coupon to get 3 pizzas for $10.

Pizzas	3	6	9	12
Cost	$10			

You can use the multiplicative nature of fractions and ratios to find the costs for 6, 9, and 12 pizzas. You don't need to figure out the cost of an individual pizza. Since 6 is twice as much as 3, 6 pizzas would cost $20 (2 × $10). Similarly, 9 pizzas would cost $30 and 12 pizzas would cost $40. This may not seem like a very complex idea, but when the numbers become more challenging, or the context more abstract, it can be very powerful.

Example

On a road trip, it took $2\frac{1}{2}$ hours to drive the first 150 miles. A ratio table can be very helpful here for estimating driving times for the rest of the trip.

Distance	150 miles	300 miles	450 miles	100 miles
Time	2.5 hours			

Since 300 is 2 × 150, we can multiply 2.5 hours by 2 to estimate the driving time for 300 miles. The same works for 450 miles, which is 3 × 150 miles: 2.5 hours × 3 = 7.5 hours.

But suppose there are only 100 miles left on the trip. It would be convenient if we knew what to multiply 150 by to get 100. Let's use a fact family:

$$150 \times ? = 100$$

$$100 \div 150 = ?$$

$$? = \frac{100}{150} = \frac{2}{3}$$

Now we know that $150 \times \frac{2}{3} = 100$, so if we multiply 2.5 by $\frac{2}{3}$, we'll get the approximate driving time for 100 miles.

$$2.5 \times \frac{2}{3} = \frac{5}{2} \times \frac{2}{3} = \frac{5}{3} = 1\frac{2}{3}$$

It will take $1\frac{2}{3}$ hours, or 1 hour, 40 minutes (since $\frac{2}{3}$ of 60 minutes is 40 minutes), to drive the last 100 miles.

Distance	150 miles	300 miles	450 miles	100 miles
Time	2.5 hours (or 2 hours, 30 minutes)	5 hours	7.5 hours (or 7 hours, 30 minutes)	1 hour, 40 min

Method #2: Equivalent Fractions

Now that we understand how ratios work, we can begin using proportions. A *proportion* is an equation that shows two equivalent ratios. For example, $\frac{3}{4} = \frac{6}{8}$ is a proportion because the fractions both represent the same part of a whole. In the above example, we can make a proportion using any of the different ratios in the table:

$$\frac{150 \text{ mi}}{2.5 \text{ hr}} = \frac{300 \text{ mi}}{5 \text{ hr}}$$

In the examples that follow, we can use what we know about equivalent fractions to solve proportions. Practicing this method is a great way to reinforce your child's number sense, understanding of fractions, and quick recall of multiplication and division facts.

Examples

1. $\frac{1}{3} = \frac{x}{12}$

We know that 3 is a factor of 12: 3 × 4 = 12. Therefore, it is simple to find out what x is:

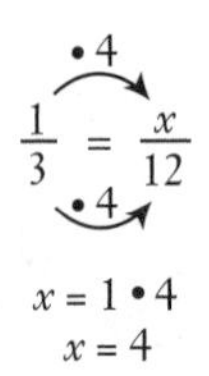

$$x = 1 \bullet 4$$
$$x = 4$$

Note: We are using the dot (•) to represent multiplication. Since x is so often used as an algebraic variable, it is preferable to avoid using × as a multiplication symbol.

2. $\frac{9}{45} = \frac{10}{x}$

In this case, 9 does not go into 10 evenly. We could multiply 9 by some fraction to get to 10, but it would be ugly, and mathematicians usually try to avoid ugly solutions because they are tedious and they invite errors. Instead, look at the relationship between 9 and 45: 9 • 5 = 45:

$$\bullet 5 \quad \frac{9}{45} = \frac{10}{x} \quad \bullet 5$$

$$x = 10 \bullet 5$$
$$x = 50$$

So $x = 50$. Note that if we reduce both of these fractions to simplest terms, we get the same thing:

$\frac{9}{45} = \frac{1}{5}$ and $\frac{10}{50} = \frac{1}{5}$

This is another way of confirming that both ratios are equivalent.

3. $\frac{6}{24} = \frac{x}{100}$

In this case, 24 does not go evenly into 100. Another problem is that x is smaller than 100. We can tell because it is the numerator, and the numerator 6 is smaller than 24. So multiplication won't work here, unless we want to involve more fractions. Let's divide instead:

$$\div 4 \quad \frac{6}{24} = \frac{x}{100} \quad \div 4$$

$$x = 100 \div 4$$
$$x = 25$$

4. $\frac{67}{4} = \frac{x}{19}$

This is about as tricky as it gets because 4 does not go evenly into either 67 or 19. You may be tempted to cross-multiply here, but let's look at it more simply. If we just had a number we could multiply or divide by, we could figure out what x is. So let's find out what 4 can be multiplied by to make 67. We can use a fact family, or the idea that multiplication and division are inverse operations:

$$4 \cdot ? = 67$$

$$67 \div 4 = ?$$

Here's a good opportunity to practice some mental math (but it's also not a bad time to use a calculator).

$$67 \div 4 = 16.75$$

Now we can use this number to solve our proportion (definitely a time to use a calculator):

$$\cdot 16.75 \quad \frac{67}{4} = \frac{x}{19} \quad \cdot 16.75$$

$$x = 19 \cdot 16.75$$
$$x = 318.25$$

The beauty of proportions is that there's always more than one way to solve them. What if we divided 19 by 4 instead of dividing 67 by 4?

$$19 \div 4 = 4.75$$

$$\cdot 4.75$$
$$\frac{67}{4} = \frac{x}{19}$$
$$\cdot 4.75$$

$$x = 67 \cdot 4.75$$
$$x = 318.25$$

Amazing!

Method #3: Unit Rates

A unit rate is a ratio that has a denominator of 1, usually expressed with the word *per*. This is often useful for both communication and calculation. If you are telling somebody about a road trip, it's great to know how fast you drove. We can simplify the ratios we found before to find out the distance traveled in 1 hour:

$$\frac{300 \text{ mi}}{5 \text{ hr}} = \frac{60 \text{ mi}}{1 \text{ hr}}$$

In other words, the travelers drove 60 miles *per* hour.

If we wanted to figure out how far they would drive in any other number of hours, it is very simple to multiply the number of hours by 60 to find their total distance. Finding a unit rate is often the quickest way to solve a proportion problem.

Examples

1. Suppose you fill your car's gas tank. You purchase 9 gallons of gas for $36. That is a ratio of $36 to 9 gallons. $\frac{36}{9} = \frac{4}{1}$, so the unit rate is $4 per gallon.

2. On a snowy day, 6 inches of snow fell over the course of 4 hours. $\frac{6}{4} = \frac{1.5}{1}$. The unit rate is 1.5 inches per hour.

3. Suppose a recipe calls for 3 cups of sugar for every 4 cups of flour, making a ratio of 3 cups of sugar to 4 cups of flour. In this case, the ratio $\frac{3}{4}$ cannot be reduced, but it can still be thought of as a unit rate. Divide both the numerator and the denominator by 4 to get $\frac{3}{4} = \frac{3/4}{1}$. That means the recipe requires a unit rate of $\frac{3}{4}$ cup of sugar per cup of flour.

Solving Problems with Unit Rates

Let's solve a few proportion problems to see how unit rates can be useful.

1. In an hour and a half, Andy washed 6 cars. How many cars can Andy wash in an 8-hour day?

   In order to solve this problem, we could use a ratio table. But since 1.5 hours does not go into the total of 8 hours evenly, it will be tricky to find the correct factor to multiply by, and the numbers might get ugly. Instead, let's find a unit rate:

$\frac{6}{1.5} = \frac{?}{1}$

$6 \div 1.5 = 4$, so the unit rate is 4 cars per hour. Therefore, to find the number of cars he can wash in 8 hours, multiply by 8:

$\frac{4 \text{ cars}}{1 \text{ hour}} = \frac{32 \text{ cars}}{8 \text{ hours}}$

Andy can wash 32 cars in 8 hours.

Perhaps the most useful thing about unit rates is that they allow you to quickly calculate several different values, as in the next example.

2. Paul read 40 pages in 1 hour and 15 minutes. How many pages can Paul read in 1 hour? 2 hours? 5 hours?

Paul's rate is 40 pages per 1.25 hours (15 minutes = $\frac{1}{4}$ hour = .25 hours). $\frac{40}{1.25} = 32$, so Paul reads 32 pages per hour. In 2 hours, he can read $32 \times 2 = 64$ pages, and in 5 hours, he can read $32 \times 5 = 160$ pages.

3. Karen scored a total of 45 points in 6 basketball games. At this rate, how many points could Karen expect to score in the whole season, which is 11 games?

45 points in 6 games is a ratio of $\frac{45}{6}$. Dividing by 6, we get a unit rate of $\frac{7.5}{1}$, or 7.5 points per game. To find the total for 11 games, multiply by 11: $\frac{7.5}{1} = \frac{82.5}{11}$, so Karen would expect to score a total of 82.5 points in the season.

The unit rate and the answer in this problem do not necessarily make sense. How could she score half a point? Rates are often used to represent an average, rather than a precise account of exactly what happened. In the first 6 games, Karen *averaged* $7\frac{1}{2}$ points per game. We have already solved word problems that involved driving for long periods of time. Drivers do not maintain a constant speed for hours at a time, but math problems are often written that way. We are meant to think of the unit rate as an average number of miles per hour, which we can use to calculate solutions to problems.

The Constant of Proportionality

Another name for a unit rate is the *constant of proportionality*.

Suppose a donut shop sells a dozen donuts for $6. That means whether you buy 12 donuts, 24 donuts, or 120 donuts, the cost per donut (or the cost per dozen) is the same. We can use a ratio table to show this:

Donuts	12	24	36	48	60
Price	$6	$12	$18	$24	$30

In this case, 1 donut would cost $0.50, so the unit rate is 50 cents per donut. As we have seen, we can use that number to find the cost for any number of donuts.

We can use the constant of proportionality to write an algebraic equation. If we let d stand for the number of donuts and P stand for the total price, then

$$P = \$0.50 \bullet d$$

This equation tells us that we can multiply the number of donuts by $0.50 to find out the total price.

Constants of proportionality will be extremely important later on in algebra, when dealing with lines and linear equations. For now, they will help us solve proportions.

Unit Conversions

Another very common use of proportions is unit conversions, where constants of proportionality are key. Suppose you take your child to the doctor, and the doctor says your child is 145 centimeters tall and weighs 42 kilograms. Most Americans would want to know what those measurements are in feet/inches and pounds. For that, we would use a conversion factor, which is another kind of constant of proportionality.

The conversion factor from kilograms to pounds is 2.2. That means that every kilogram is equivalent to 2.2 pounds. If a child weighs 42 kilograms, how many pounds is that? Every kilogram is 2.2 pounds, so we can multiply $2.2 \times 42 = 92.4$ pounds.

Similarly, every inch is 2.5 centimeters. So if a child is 145 centimeters tall, we need to solve the proportion

$$\frac{145 \text{ cm}}{x \text{ in}} = \frac{2.5 \text{ cm}}{1 \text{ in}}$$

$145 \div 2.5 = 58$, so multiply 1 inch by 58, and the child is 58 inches (4 feet, 10 inches) tall.

Method #4: Proportional Equations

Mathematicians love shortcuts, abbreviations, shorthand—in other words, whatever makes things more concise and efficient. The more information they can pack into a few symbols, the better. This is one of the reasons many people find math so confusing—because it sometimes takes a lot of understanding to unpack the meaning crammed into a tiny equation like $A = \pi r^2$.

The equation that is used to represent proportional relationships is very simple. It is taught early because it sets students up for future work in algebra.

The Equation for a Proportional Relationship

Any proportional relationship can be summarized by the equation

$$y = kx$$

Note: When two variables are side-by-side, or a number is directly in front of a variable, it implies multiplication. So kx means *k times x*.

There are many different forms this equation can take. You may see $m = dv$ or $d = rt$. The specific letters (called *variables* because they represent a quantity that can change, or vary) are not important—what is important is that you understand the purpose of each letter.

In the equation $y = kx$:

- x and y represent two quantities that have a proportional relationship to one another.
- k is the constant of proportionality between x and y.

Examples

1. Let's take one of the unit conversions above as our first example. We saw that every kilogram is equal to 2.2 pounds. Therefore, kilograms and pounds are proportional to each other, with 2.2

being the constant of proportionality. We can rewrite the general equation as

$$\text{pounds} = 2.2 \times \text{kilograms}$$

or simply as

$$y = 2.2x$$

where x is the number of kilograms, and y is the number of pounds.

Using this equation, it is simple to find the number of pounds that are equivalent to any given number of kilograms. Suppose something has a mass of 50 kg.

pounds = 2.2 × kilograms
pounds = 2.2 × 50
pounds = 110

So 50 kg = 110 pounds.

OTHER FORMS

Sometimes it is preferable to change the variables in a problem. Rather than always using x and y, you may find that other letters are more useful in a given context. In the example above, you could use p for pounds and k for kilograms, making the equation

$$p = 2.2k$$

Suppose you are solving a problem about a bicyclist riding at a constant speed. In this case, it would make sense to use the equation $d = rt$. The variables have meaning—d means distance, r means rate, and t means time. If a person is moving at a constant speed (r, your constant of proportionality), the distance traveled is proportional to the time spent traveling.

2. Suppose a bicyclist is riding at a constant speed of 18 miles per hour. The amount of time she spends cycling will determine the distance she travels, and the constant of proportionality is 18 miles per hour. Therefore, the equation $d = 18t$ describes the relationship between her distance traveled and her time spent. If we choose any number of hours and replace t with that number, we can find the distance by multiplying 18 by t.

Which Method, and When?

We have studied four methods for solving proportions: ratio tables, equivalent fractions, unit rates, and equations. What all of these methods have in common is strategic multiplication and/or division. In this way, they are no different from cross-multiplication, and as you have worked through the examples in this chapter, you may have noticed the connection to that method.

Not every method is going to make sense to every student (or parent). The important thing is to find one method that really works for you and master it, while still trying to make sense of the other methods. Some methods are better for certain situations than others, and good mathematicians are efficient because they have access to a variety of methods—just as a good chef can choose between broiling, roasting, grilling, etc., depending on what will give the best result for the dish. This is the incredible power of the new teaching methods—students now are empowered to choose methods that *make sense* to them, while they are also taught the flexibility they need to adapt to a variety of challenges.

Percentages: A Special and Very Common Use of Proportions

A *percentage* is a ratio with a denominator of 100. When a percentage is used to describe quantities, the whole is always thought of as 100%.

Finding a Percentage

Since a percentage is a ratio out of 100, any ratio can be converted to a percentage—some more easily than others. Let's return to an example from early in this chapter. A class of 25 students has 10 boys and 15 girls. We can figure out what percentage of the class is made up of boys:

Part	10 boys	?
Whole	25 students	100

Since $25 \times 4 = 100$, multiply 10×4 to get 40. Boys make up 40% of this class.

Using a Percentage to Find a Part

Newspapers often poll groups of people and report the results in percentages. In one poll, a newspaper asked 2,000 people their preference in an election. 60% of respondents said they will vote for Candidate A. You can figure out exactly how many people that is based on the number of people who participated in the survey. Let's use a ratio table. Remember that 60% means $\frac{60}{100}$.

Part	60	x
Whole	100	2,000

Since $100 \times 20 = 2{,}000$, we can multiply 60×20 to find out how many people said they will vote for Candidate A: $60 \times 20 = 1{,}200$. We will work with polls and statistics more in Chapter 10.

Using a Percentage to Find the Whole

You can also use percentages to find the whole when you know a part. For example, suppose a pair of pants costs $45 after a 25% discount. You want to figure out the original price of the pants. We don't know how much the 25% discount was, but we know the final price *after* the discount. Since the original price was 100%, and we discounted the price by 25%, the final price is 75% of the original price. Therefore we use 75% in our ratio table:

Part	75	$45
Whole	100	x

You could try to figure out what number to divide 75 by in order to get 45: $75 \div 45 = 0.6$, so $75 \times 0.6 = 45$, and $100 \times 0.6 = x$. That's one way. But there are other, more elegant methods.

We can use the equivalent fractions method to solve this problem in two steps. First we reduce $\frac{75}{100}$:

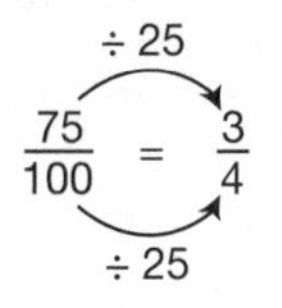

Then, we can multiply to get a new fraction with a numerator of 45:

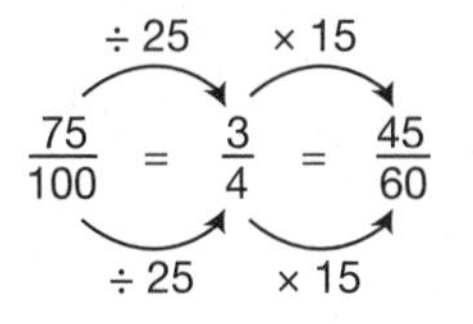

So if the part is 45, then the whole is 60. The original price of the pants was $60.

Sales Tax

Whenever you go shopping, items rarely end up costing what the sticker says—an item labeled $0.99 ends up costing more than a dollar. Suppose sales tax in an area is 8%. That means that for every $100 you spend buying items, you will have to pay an additional $8 in tax.

Since a percentage is a ratio out of 100, 8% means $\frac{8}{100}$, or 0.08 in decimal form. If you want to calculate the tax on an item, you can multiply the price of the item by 0.08.

> **A COMMON ERROR**
>
> When dealing with single-digit percentages, many students will make a mistake with place value. When converting 8% to decimal form, students (and adults) often write 0.8, rather than 0.08. Similarly, many students represent a nickel as $0.5, which is really half of a dollar, or 50 cents. If your child makes this error, remind him or her that 8% is 8 out of *100*, which is 8 *hundredths*, not 8 tenths.

Examples

For the following examples, let's keep using an 8% tax rate. Calculating sales tax is a great time to use an equation. There are two common ways of finding the total price.

1. A sweater costs \$25. What is the total cost, after tax?

Let's solve this problem by first figuring out the tax, then adding it to the original price. Since the tax rate is 8%, we can use 0.08 as a constant of proportionality. Using t to represent the tax and p to represent the price, our equation is

$$t = 0.08p$$

The sweater costs \$25, so the tax is

$$t = 0.08 \cdot 25$$
$$t = 2$$

So the tax is \$2, and the total cost of the sweater after tax is \$25 + \$2 = \$27.

We solved this problem in two steps: first finding the tax using 0.08 as a constant of proportionality, then adding the tax to the original price. We can also find the total price in one step, if we use 1.08 as the constant of proportionality. Why 1.08? Because when you pay 8% tax on an item, you are really paying a *total* of 108% of the price of the item—100% for the item itself, and 8% for the tax. We will use this method on the next example.

2. A lamp is on sale for \$53. What is the total cost, after tax?

For this one, we can let T represent the total cost, and p represent the price of the lamp.

$$T = 1.08p$$
$$T = 1.08 \cdot 53$$
$$T = 57.24$$

So the total price of the lamp after tax is \$57.24.

PRACTICE MATH ON THE GO

States and cities each have their own sales tax rates. If you don't know your local sales tax rate, find it out. When shopping, ask your child to estimate the total cost of items after tax.

Tips

Being able to calculate a tip is even more important than being able to calculate tax, because cash registers and other machines usually figure out the tax for you—the government and the businesses don't trust anybody to calculate tax on their own. Fortunately, there is a good amount of wiggle room when it comes to tips. 15–20% is a standard tip in a restaurant, but most people don't pull out a calculator to make sure they leave that amount to the penny. So good estimation skills come in handy.

Examples

1. How much would a 20% tip be on a $28 restaurant bill?

20% is a fairly simple percentage to work with. Notice that 10% of any number is $\frac{1}{10}$ of that number, since 10% = $\frac{10}{100}$, which simplifies to $\frac{1}{10}$. It's easy to find $\frac{1}{10}$ of a number—just move the decimal point one space to the left (changing the place value of every digit in the number by one place, as you learned in Chapter 3). So 10% of $28 is $2.80, and 20% is $5.60. $5 or $6 would be an appropriate tip.

2. How much would a 15% tip be on a $41 check?

If the bill is $41, let's just round it down to $40. Calculating 0.15 × 40 is not too difficult if you break the .15 into its place value components. First, find that 0.1 × 40 = 4. Then, 0.05 is half of 0.1, so 0.05 × 40 is 2. The tip would be around $6 ($4 + $2). You may also know that 15 × 4 = 60, since 15 minutes is a quarter of an hour, so 0.15 × $40 = $6.

A QUICK TRICK

Look at the following fraction-decimal equivalences:

$$\frac{1}{5} = 0.2$$

$$\frac{1}{6} = 0.1\overline{6}$$

Since $\frac{1}{5}$ = 20%, you can easily calculate a 20% tip by dividing the total by 5.

If you want to estimate a 15% tip, divide the total by 6, since $\frac{1}{6}$ is pretty close to 15%.

Summary

Proportions are one of the most important topics in middle school math. Not only are they an essential part of the foundation for future work in algebra and geometry, but they are guaranteed to pop up in your day-to-day experiences. Ask your child about ratios in his life. What is the ratio of boys to girls in your math class? What is the ratio of adults to kids at a family gathering? The examples throughout this chapter should give you an idea of some of the places in life where you can find ratios and proportions. Challenge your child to see who can find more ratios in a day. Then turn each ratio into a proportion problem. If you spend $30 on 7 gallons of gas, ask your child how much 14 gallons would cost, or what the unit rate is per gallon.

Proportions are about a lot more than just the old-fashioned cross-multiplication algorithm. In this chapter, we have used four different methods that your child's teacher will likely use in class:

- Using ratio tables to find and follow patterns
- Converting between equivalent fractions
- Finding and using unit rates
- Equations that use the constant of proportionality ($y = kx$)

The purpose of providing so many different methods is to give your child a deep understanding of the concept and flexibility in problem solving. If you were skeptical about the new teaching methods before, hopefully this chapter has helped you see how your child can benefit from a more conceptual approach. But the old ways certainly had their positives, and an emphasis on practice was one of them. The more your child practices with different methods, the better he will understand any concept he studies.

The work we have done here with proportions as an algebraic concept will be important later on—particularly in Chapter 8, when we tackle the infamous "solve for x" problems.

In Chapter 5, we will turn our attention to another important feature of modern math instruction: visual representations. As you will see, the more you can visualize a math concept, the better your understanding of it will be. Chapter 5 will introduce some basic geometry concepts and use them as representations for other aspects of mathematics.

Visualize Everything

How Simple Geometry Concepts Can Make Numbers, Operations, and Algebra Easier

The Power of Visualization

Visualization is one of the great features of a modern math classroom. Good math teachers use visual models for nearly any concept, giving students access to more ways of understanding new ideas. In many old-fashioned math classrooms, visualization was rarely used. Fraction operations were taught as a step-by-step procedure, without ever showing what the fractions look like. Some teachers even taught *geometry* without ever drawing a picture of the shapes they were talking about! This worked for some students, but many students were left behind.

While these kinds of pictures may seem "elementary" to some, it truly makes a difference for students and adults alike. The term *visual learner* gets used a lot in education to describe students who (supposedly) can only learn or master a concept if they can see it. To some extent, all students are visual learners. When a student is stuck on a word problem, fraction operation, or new algebra concept, often the quickest way to help her through it is by drawing a picture.

In previous chapters, we have already used visual models to represent integers and fractions to help us understand why the four operations (addition, subtraction, multiplication, and division) work the way they do on these types of numbers. Now we will continue to develop that understanding as we delve into the basics of geometry—length and area measurements.

The Three Dimensions

Early in middle school, most of your child's work in geometry will revolve around the concepts of length and area. To understand the difference, it is helpful to be able to distinguish between one, two, and three dimensions. If you or your child has ever seen a 3-D movie, then you already have some understanding of what the three dimensions are. What makes 3-D movies so incredible is the way the action seems to pop out of the screen, right in front of your eyes. These images are not limited to the flat, 2-D world of the movie screen.

A **two-dimensional** object is something flat, like this page, or a movie screen. We call these items two-dimensional because they can be measured in two different ways. A painting can be measured by its length (top to bottom) and its width (left to right). While technically the thickness of the painting is nonzero and can also be measured, paintings, paper, and screens are good models for understanding two-dimensional objects.

Three-dimensional objects are all around us. Anything you can pick up and hold in two hands is three-dimensional, even if it is very thin. A shoebox is a great example of a three-dimensional object because you can easily see its three dimensions. It could be measured along its length (front to back), its width (left to right), and its height (top to bottom).

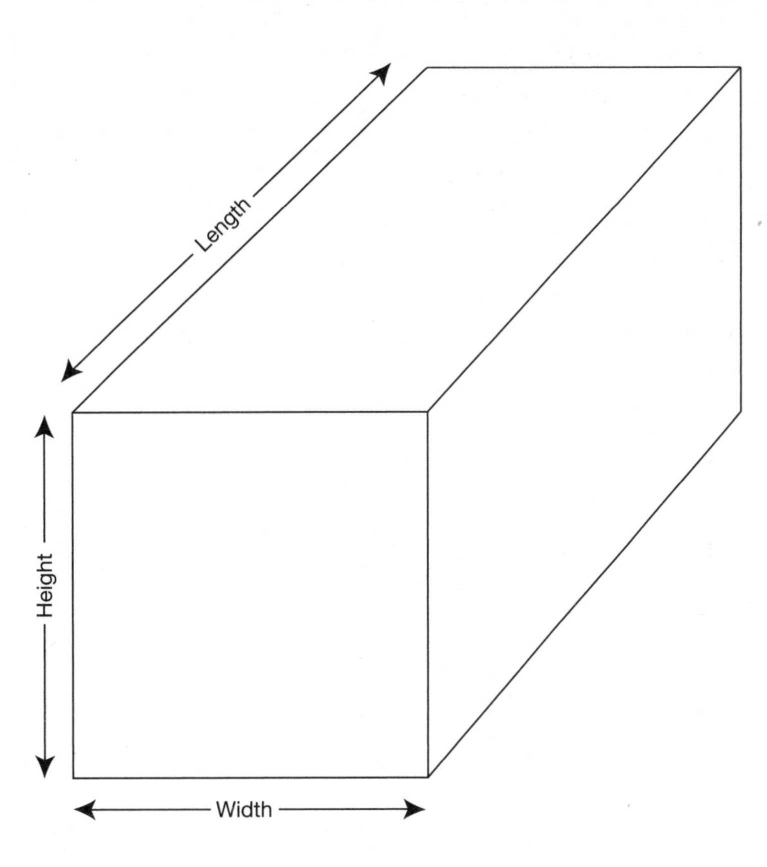

For our purposes, we will focus on two-dimensional geometry and its usefulness as a visualization tool.

Length and Area

We are most familiar with measurements of length, which can be any one-dimensional measurement. Your child's height, the distance to school, and the distance across a room are all examples of length measurements. When describing a length, we always use a unit such as inches, feet, yards, or miles in the English system or millimeters, centimeters, meters, or kilometers in the metric system.

When measuring a two- or three-dimensional object, you usually start by measuring each of its dimensions. While the dimensions of a box are labeled as **length**, **width**, and **height**, it is important to know that all three of these dimensions are actually measurements of length, meaning that they are one-dimensional measurements. The width of a box is a length, and so is its height.

Area is a difficult concept for many students to understand at first. When renting or buying a home, you would likely be interested in its square footage. This is another way of referring to the area of the living space. A room cannot be measured in inches or feet because such a measurement would fail to describe how much open space is in the room (it could only tell you how long one side is). We need two-dimensional units to describe an area, so we use **square units**, such as a square foot. A square foot is the area of a square that is one foot long and one foot wide.

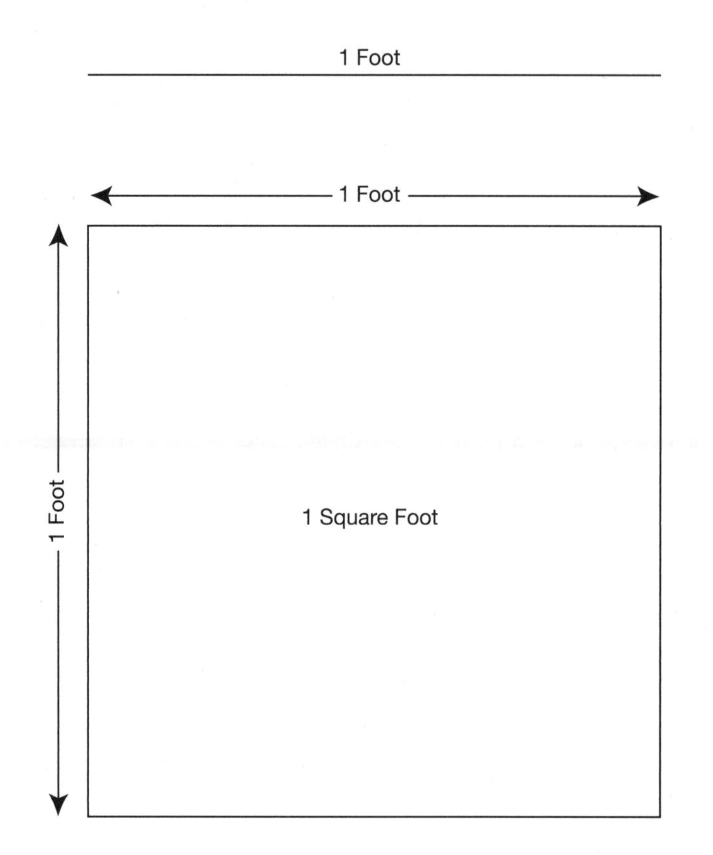

So if you live in a home that is 1,500 square feet, then it means you could cover the floor of the entire home with 1,500 one-foot-by-one-foot tiles. Some common units of area include square inches, square centimeters, and square miles.

Finding the Area of a Polygon

A polygon is a closed two-dimensional shape with one interior (inside area), whose exterior is made of straight line segments. As often happens in math, that's a very simple definition made up of a lot of complicated words. It's the nature of mathematicians—we love to compress as much meaning into as few words as possible. This definition is a great example of *how* terms are defined in math—it tells you exactly what the thing *is*, as well as what it *is not*.

Let's unpack that definition. A polygon is a *closed two-dimensional shape*. That means it is flat (not three-dimensional), and it does not have any openings. If an ant were inside, there would be no way out.

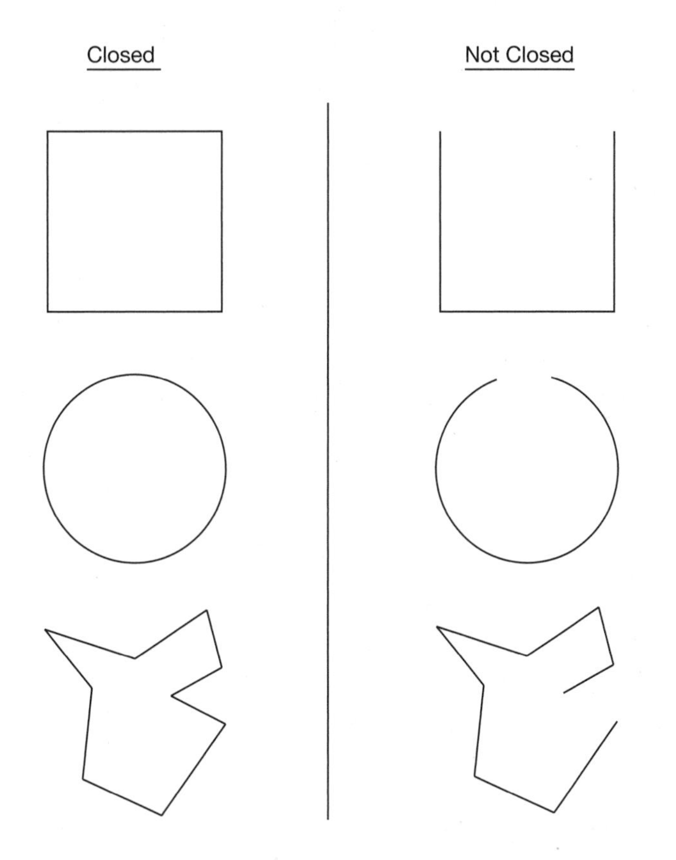

A polygon has *one interior*. That means that if two ants were trapped inside the shape, they would be able to find their way to each other, no matter where they were inside. Shapes with more than one interior would have separate compartments where the ants could get stuck.

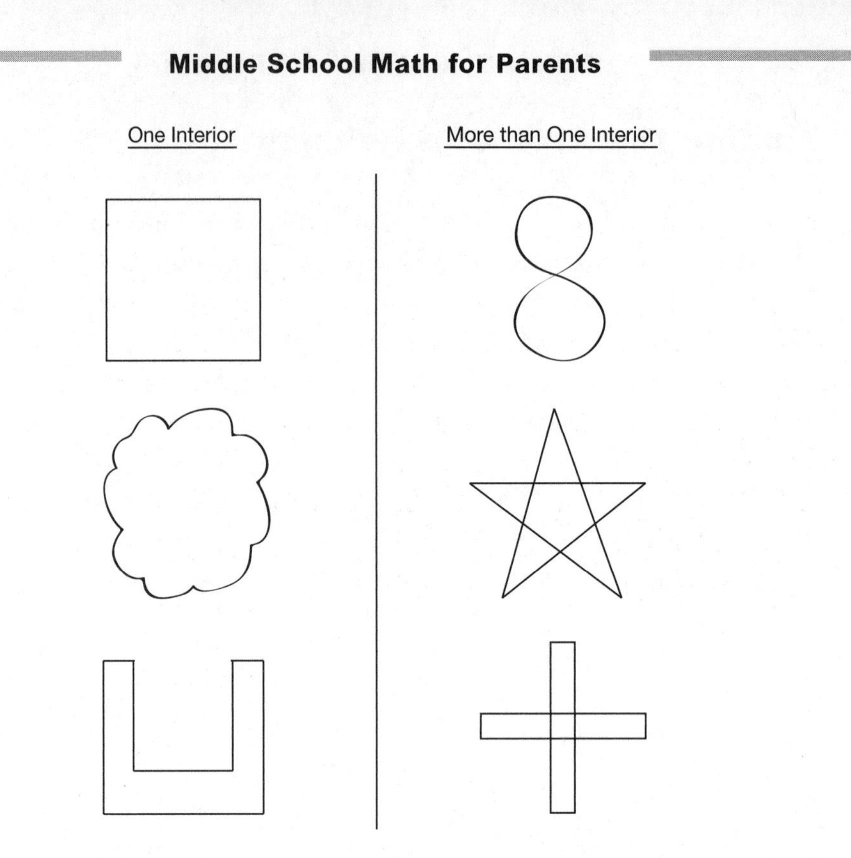

Finally, a polygon's exterior must be *made up of straight line segments*. That means no curves. The reason we say *line segments* instead of *lines* is because the mathematical definition of a line says that it goes on forever in both directions. In a polygon, the "lines" end at the corners of the shape.

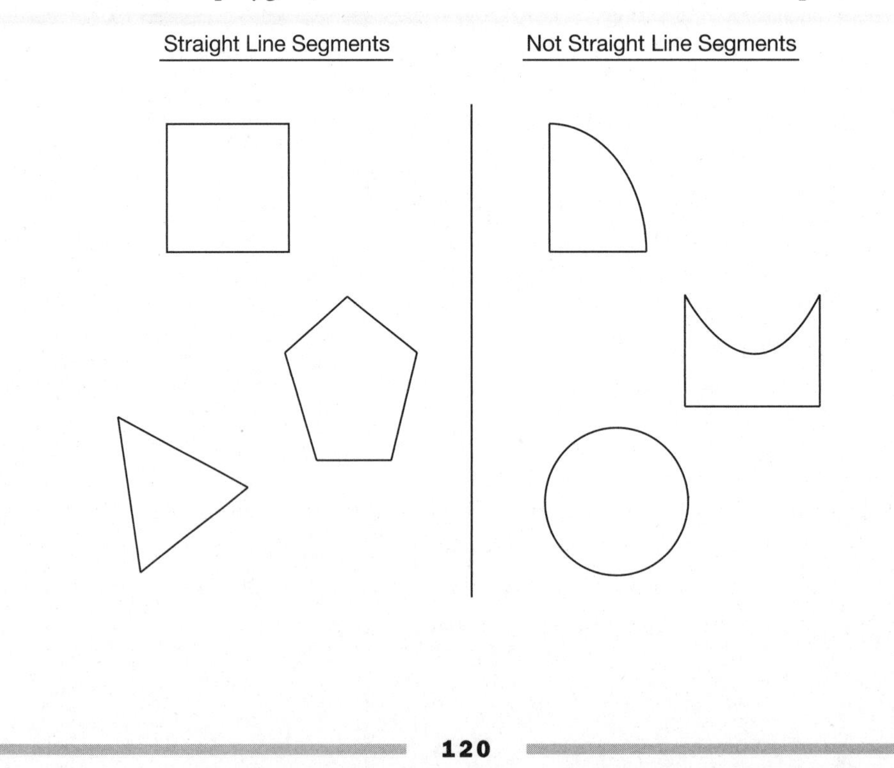

With these bits and pieces of the definition understood, we can make up lots of examples and non-examples of polygons. You and your child can practice applying the definition of a polygon by naming the reason why each non-example is not a polygon.

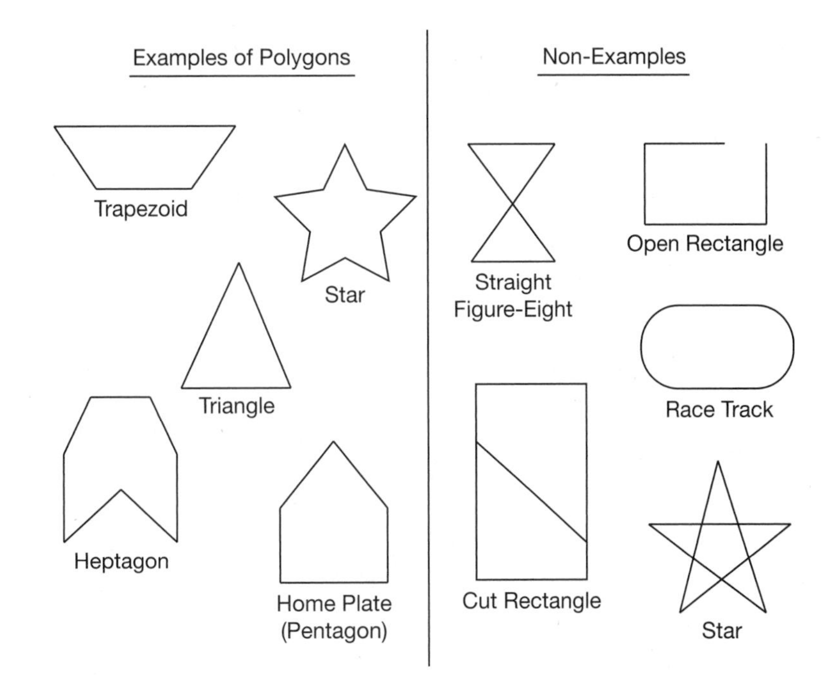

IDEA FOR PRACTICE

Making lists of examples and non-examples is a great way to build and test your child's understanding of any new math term.

Area of a Rectangle

Your child likely learned to find the area of squares and rectangles in elementary school. If you draw any rectangle on a grid, the area of the rectangle is the number of square units that fit inside it. You can think of each grid box as a one-unit by one-unit square, which is one square unit. Math problems will often use generic "units," rather than naming a specific unit like inches.

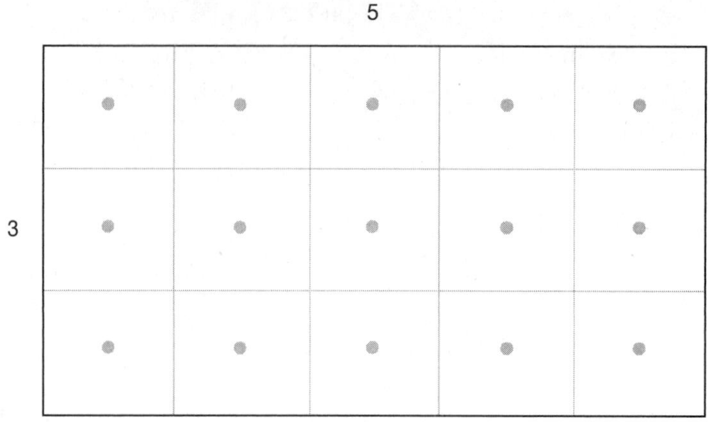

5

3

In this example, we can simply count the number of square units inside the shape—the area is 15 square units. But your child will probably recognize very quickly that the rectangle has 3 rows of 5 square units each, so multiplying 3×5 would give the area faster than counting.

If you multiply the length of a rectangle by the width, the product is the area:

$$A = l \cdot w$$

SQUARES AND RECTANGLES

Many students misunderstand the relationship between squares and rectangles. The definition of a rectangle is *a four-sided shape with four right angles*. Therefore, every square is considered a rectangle, since its corners are all right angles. The misconception that a rectangle needs to have two different side lengths is very common—an idea that develops early on from the way shapes are presented to young children. Ask your child if a square is a rectangle, and help clarify it if he or she is not sure.

A simple analogy is this: every woman is a human, but not every human is a woman. Every square is a rectangle, but not every rectangle is a square. *Square* is a more specific word to identify *certain* rectangles. *Woman* is a more specific word to identify *certain* humans.

Using Area as a Visual Tool

Since the area of a rectangle is the product of the rectangle's length and the width, we can use a rectangle to represent *any* multiplication problem. What's 6 × 19? It's the area of a rectangle 6 units long and 19 units wide. As it turns out, this idea can be incredibly powerful when multiplying multi-digit numbers, or even algebraic expressions.

In Chapter 3, we used decomposition of numbers as a strategy for multiplying multi-digit numbers. We were able to solve a problem like 6 × 19 by multiplying 6 × 10 and 6 × 9, then adding the products together. But we saw that when it came to *two* two-digit numbers, things got complicated. 18 × 57 turned out to be a bit too hard for a quick calculation.

Using the area of a rectangle is a great way to visualize a problem like this. When we break each number into its place value components, 18 becomes 10 + 8 and 57 becomes 50 + 7. An 18-by-57 rectangle can be broken up into four sections:

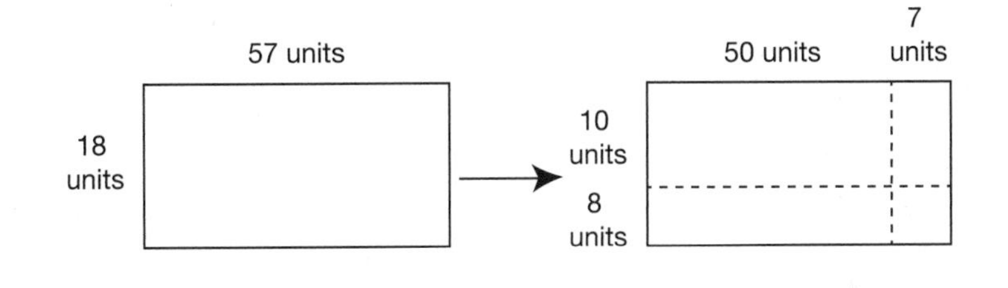

Now it is very easy to find the area of each of the four sections, using our multiplication facts and our knowledge of place value.

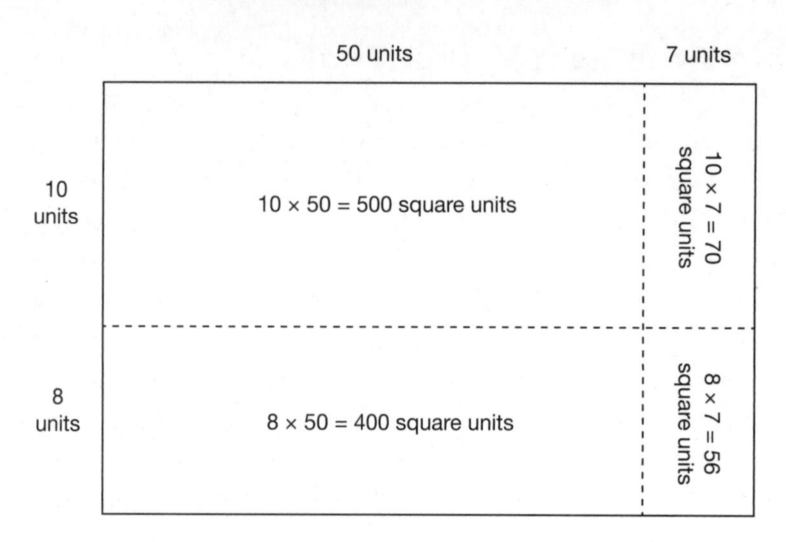

To find the product of 18 and 57, all we need to do is add up the four areas, giving us the total area of the rectangle: 500 + 400 + 70 + 56 = 1,026. Therefore 18 × 57 = 1,026.

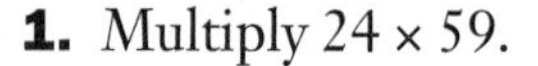

Examples

1. Multiply 24 × 59.

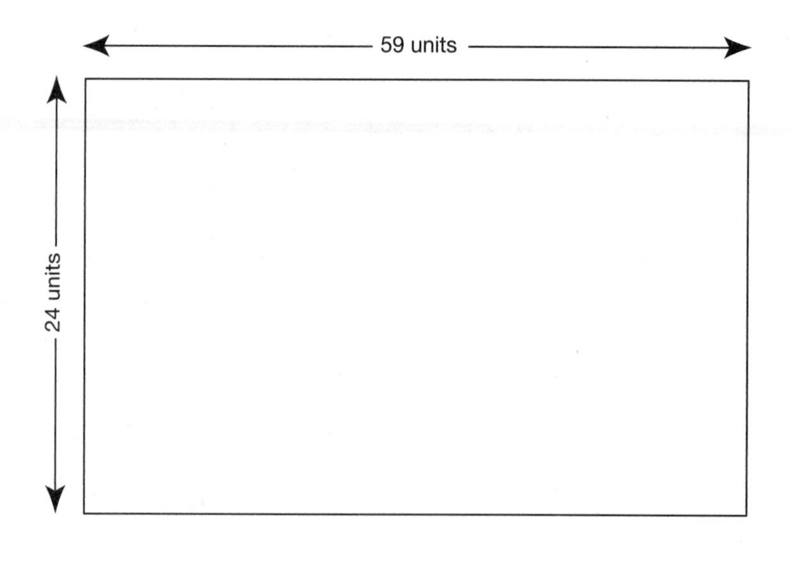

We could break each number into its place-value components, just as we did for 18 × 57. Or we could use an area model where we only break up one number, say the 59. When we break the 59 up into 50 + 9, our rectangle looks like this:

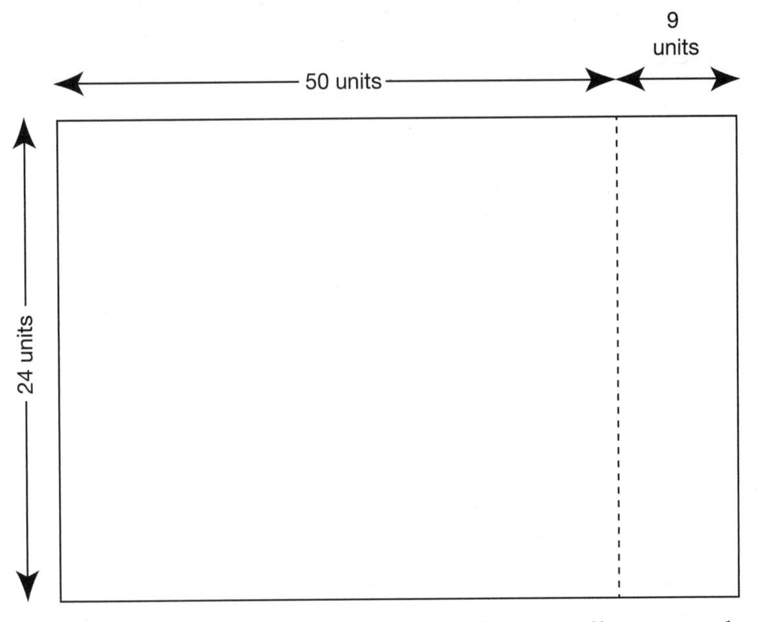

The original rectangle is made up of two smaller rectangles: a 24 × 50 and a 24 × 9.

24 × 50 = 1,200 and 24 × 9 = 216, so the total area is 1,200 + 216 = 1,416 square units.

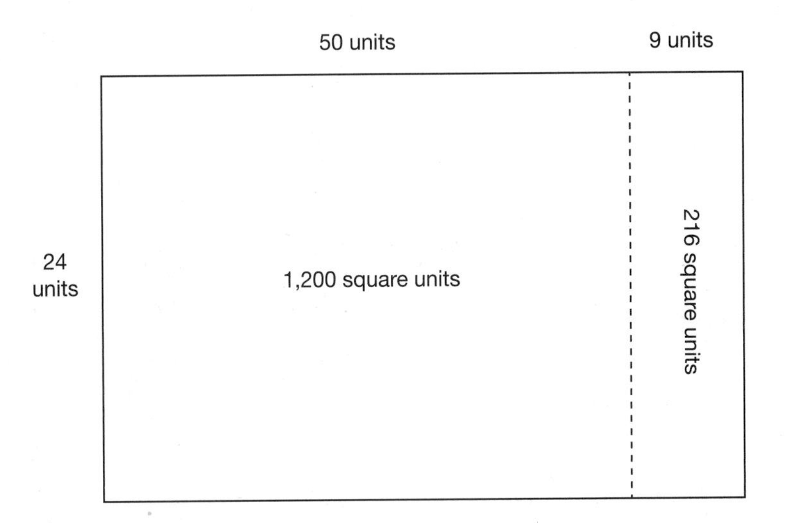

The beauty of this method is that you are free to choose how much you need to break up the numbers. If you have trouble multiplying 24 × 9 in your head, you can break the 24 up further to make the work easier.

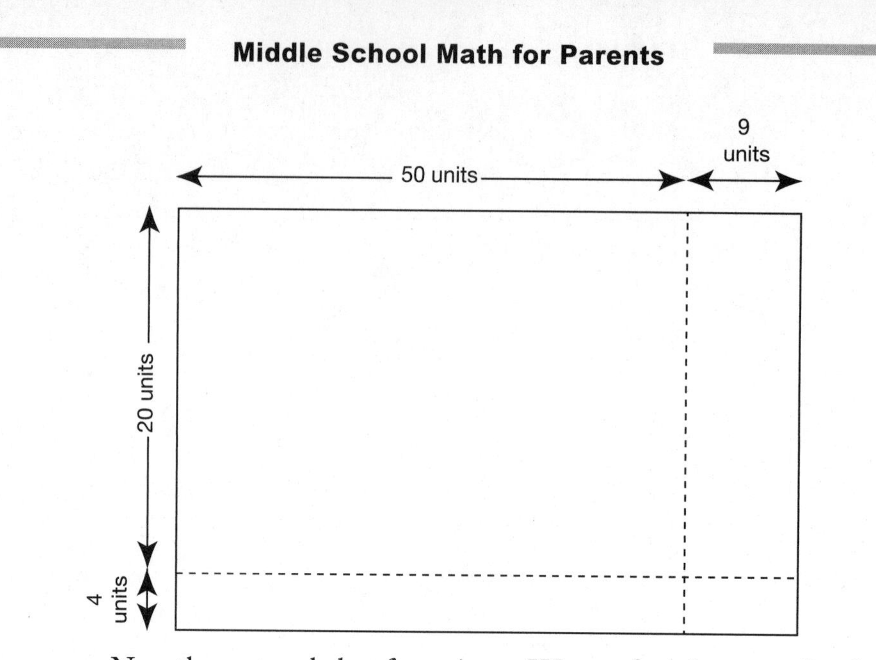

Now the rectangle has four pieces. We can find the area of each piece, and add them all up:

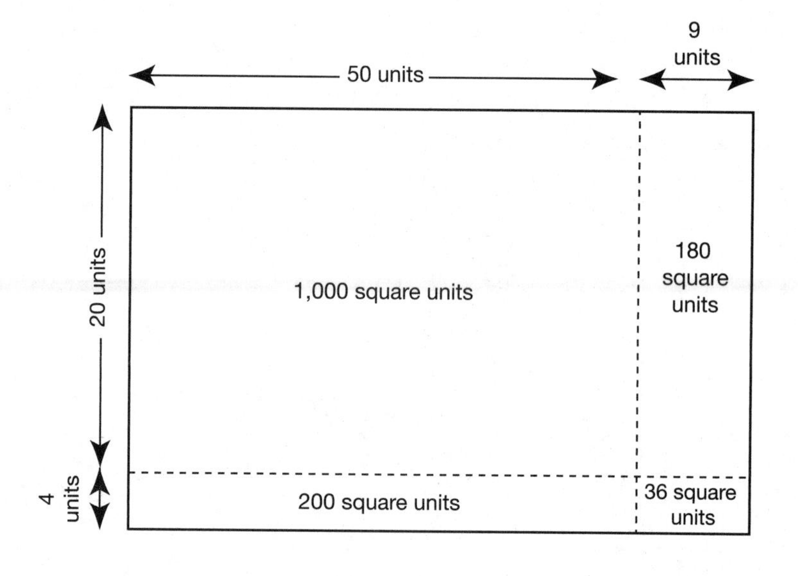

$20 \times 50 = 1{,}000$
$20 \times 9 = 180$
$4 \times 50 = 200$
$4 \times 9 = 36$
$1{,}000 + 180 + 200 + 36 = 1{,}416$

By showing a visual representation of what it means to decompose two factors, you can help your child understand the method of decomposition better. The power in this is that many students find it simple to perform this visualization method mentally.

We can also use this strategy for numbers with three or more digits.

2. 147 × 392

In this case, we can break each number into its hundreds, its tens, and its ones. Don't worry too much about making sure everything is the right size—it would take up way too much paper to have your hundreds actually look 100 times as large as your ones.

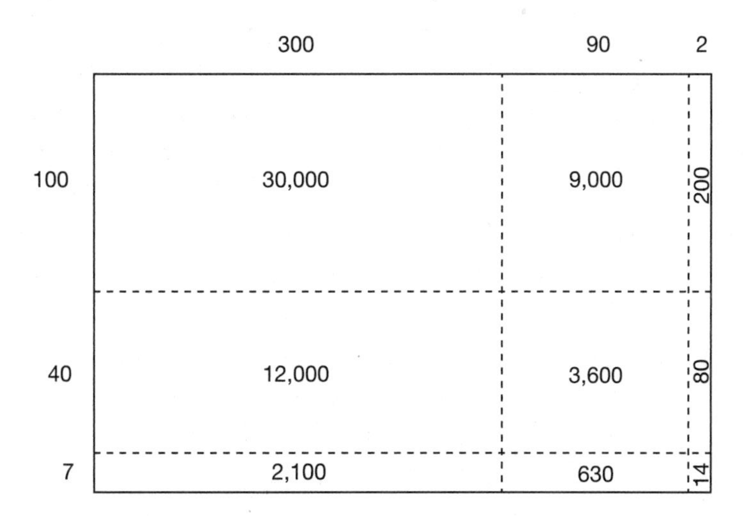

There are 9 sections in this rectangle because each of our factors had 3 digits (3 × 3 = 9). That's a lot of areas to add up, but for most students, it beats the old error-prone algorithm. We could add up each column (or row), and then add them all together:

30,000 + 12,000 + 2,100 = 44,100
9,000 + 3,600 + 630 = 13,230
200 + 80 + 14 = 294
And the grand total: 44,100 + 13,230 + 294 = 57,624.

In case you are worried, this is not a problem I would recommend solving mentally. In reality, it's a great time to use a calculator. But the point of using the area model is to help your child make the connections between area and multiplication.

The same strategy works for decimals, as in the following example. Again, we aren't going to try too hard to make the picture the correct size—it's the numbers that are important.

3. Multiply 4.5×0.75

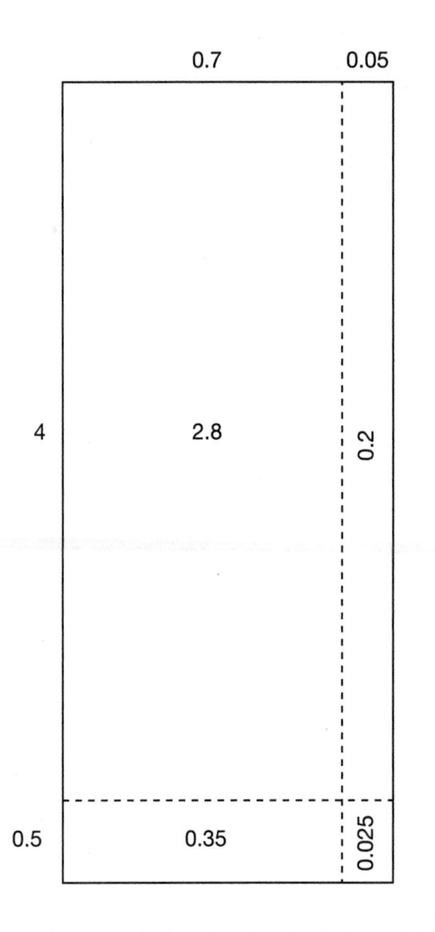

The total area is $2.8 + 0.2 + 0.35 + 0.025 = 3.375$.

We can also solve the last example in fraction form. Since 4.5 = $4\frac{1}{2}$, and 0.75 = $\frac{3}{4}$, we could draw our area model this way:

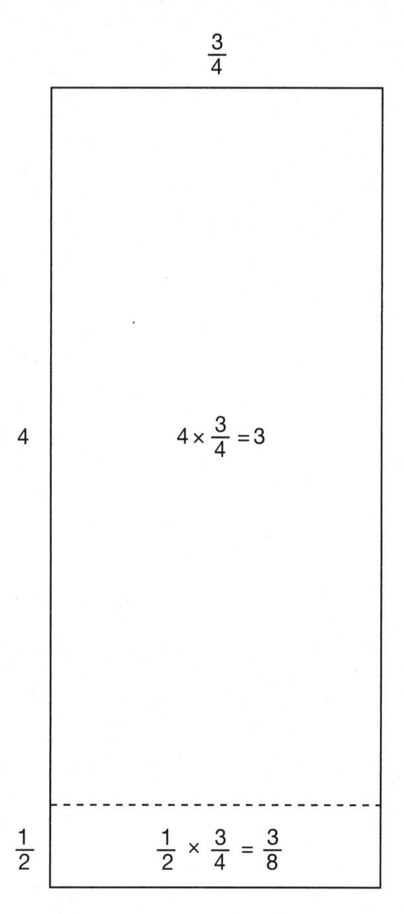

So the total area is $3\frac{3}{8}$, which is equivalent to 3.375.

An Algebraic Example

When your child studies algebra, he or she will learn about the **Distributive Property**. You may remember the mnemonic **FOIL**. While we aren't going to solve these types of problems until we get to algebra, one algebraic example will illustrate the usefulness of area as a visualization technique for algebra problems.

We could multiply $(x + 5)(2x + 6)$ this way:

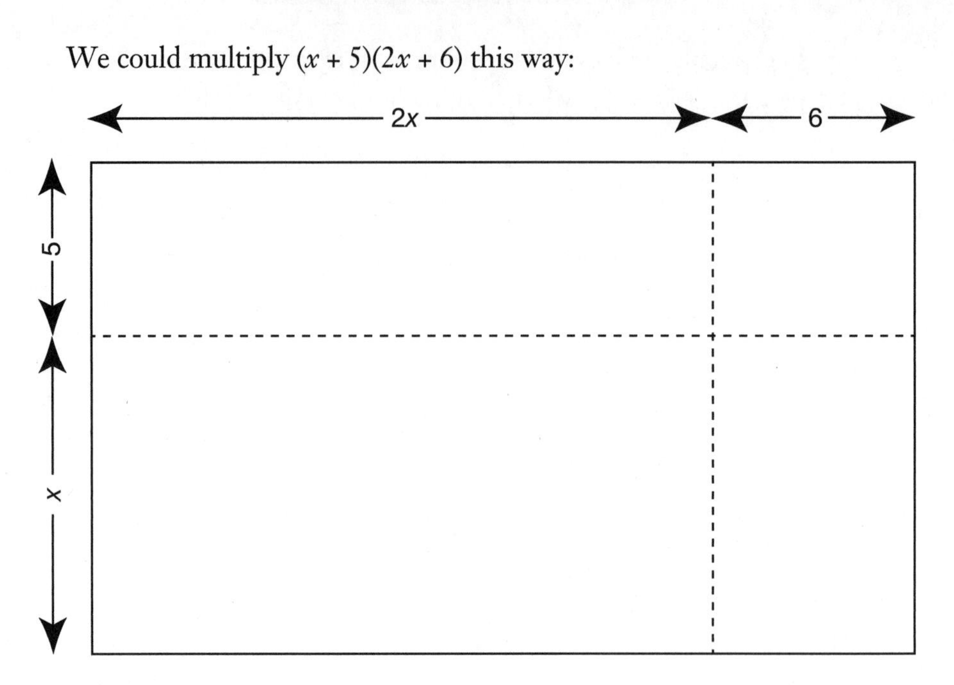

No need to worry if this algebraic example doesn't make much sense to you right now because we will come back to this concept in Chapter 7. For now, the important thing is to know that it's coming—that's why it's so important for your child to understand now how area can be used to visualize *any* multiplication problem.

Area of a Triangle

You may remember the formula $A = \frac{1}{2}bh$. Years ago, most teachers gave students this formula, and then asked them to solve many examples, often without providing a drawing. These days, students have the benefit of a more common sense, visual approach to geometry. We will examine where this formula comes from, beginning with right triangles.

Right Triangles

A **right triangle** is any triangle that has one 90° angle.

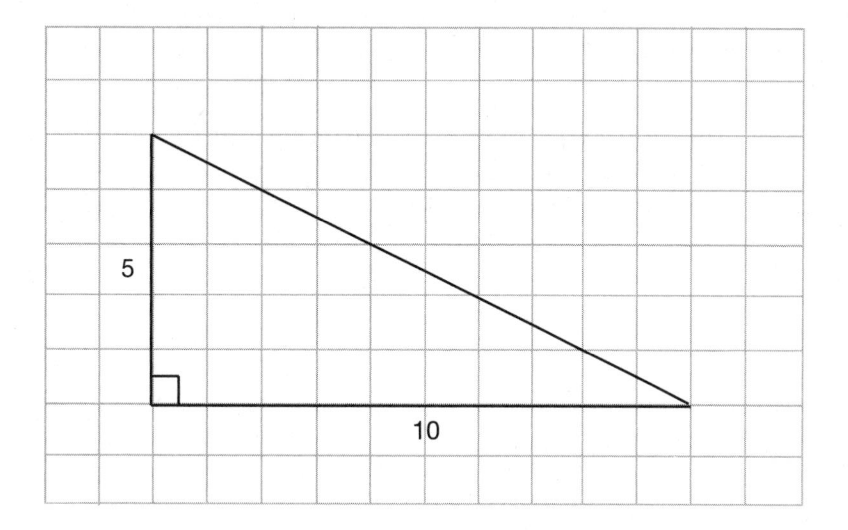

This triangle has a height of 5 units, and a base (width) of 10 units. It would be difficult to count the squares that make up the area of this triangle because so many of them are chopped into oddly shaped pieces.

A COMMON ERROR

Many students try to find the area of this triangle by counting boxes. Some count any box that the diagonal side passes through, while others count each of those as half of a box, since they are all cut into pieces. While some teachers may show students tricks for fitting those chopped-up pieces together, it is important to understand that this is an estimation method, not a precise method for calculating the area of a triangle.

There is a more precise method. If we duplicate the triangles, and arrange the two copies just right, the area is simple to find:

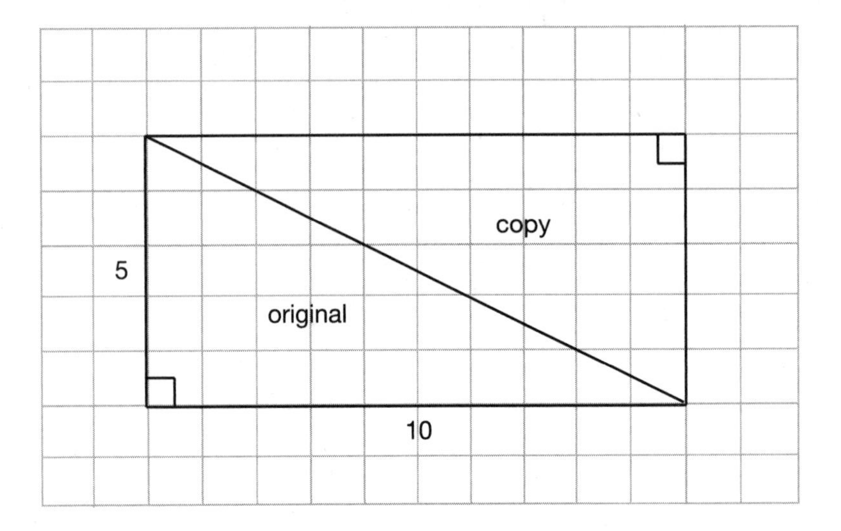

Any right triangle is half of a rectangle that is cut diagonally corner-to-corner. So by copying the triangle, you can form the full rectangle. The area of this rectangle is 5 units × 10 units = 50 square units. So the area of each right triangle is half that amount, or 25 square units. Pretty cool, huh? If your teachers never showed you this method, hopefully you can appreciate the power of visualization.

Other Triangles

So we see now why the $A = \frac{1}{2}bh$ formula works for right triangles. But what about other kinds of triangles, since not every triangle has a 90° angle? Consider the following example.

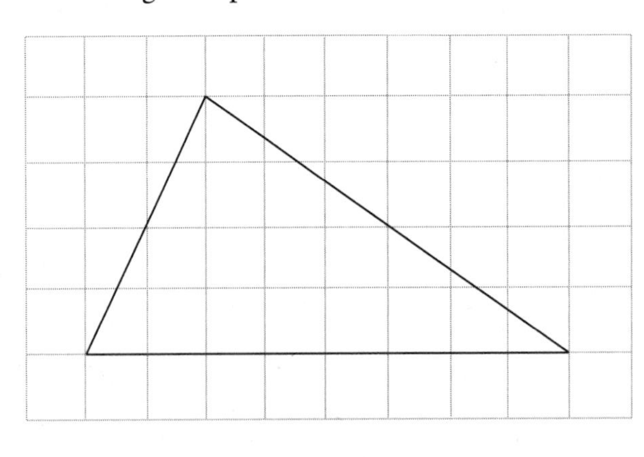

There is no straightforward way to fit this triangle into a rectangle, but there's always a trick. What we can do is slice the triangle vertically into two triangles. In my classroom, we refer to this as using our "ninja skills." Many students are pleasantly surprised to learn that ninjas are mathematicians, too.

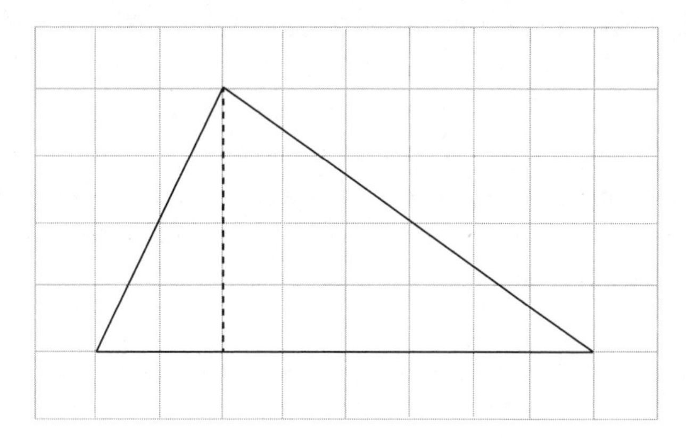

Now you can see that both of the new pieces are right triangles. We could copy each of those triangles to fill out a rectangle for each:

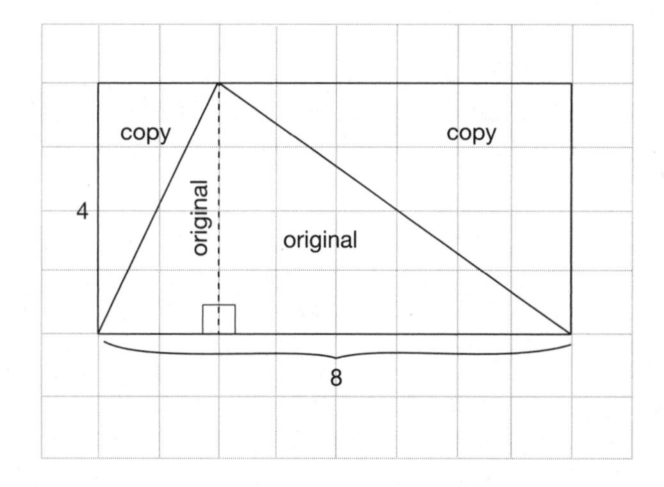

We have now fit our original triangle and its copy into a rectangle. The area of the rectangle is double the area of the original triangle. The rectangle is 8 × 4—its area is 32 square units. Therefore, the area of the original triangle is 16 square units. Notice that the 8 is the length of the base of the triangle, and the 4 is the height of the triangle. Our old friend $A = \frac{1}{2}bh$ has not failed us, and we have now seen that this method works for *any* triangle—not just right triangles.

Area of Other Polygons

You may remember memorizing formulas for the areas of different kinds of shapes. The areas of parallelograms and trapezoids are notable examples of formulas that many of us have memorized, but few are likely to remember. For your child, life will be a little easier than that. Knowing how to find the area of a rectangle and a triangle will allow your child to find the area of any polygon, no matter how crazy. The trick is thinking of *any* polygon as a **composite figure**—that is, a figure that is made up of smaller, simpler figures. Since a polygon always has straight sides, every polygon can be cut into triangles and rectangles.

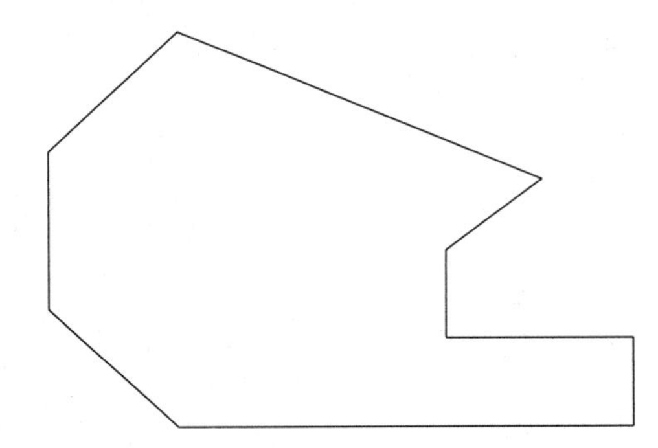

This polygon looks really complicated! Here is the same shape again, shown as a composite of rectangles and right triangles:

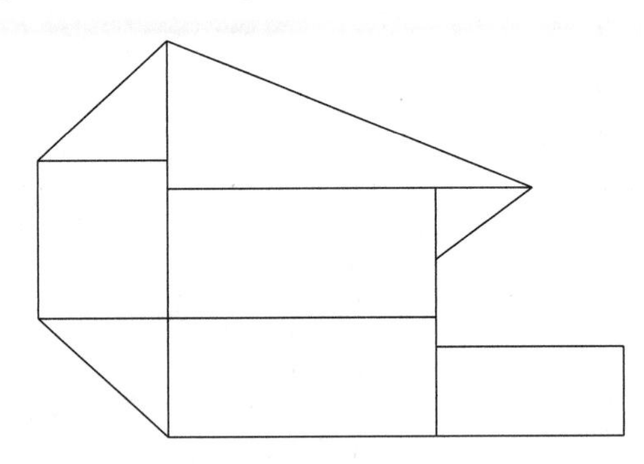

It will take some work to find the areas of all those rectangles and triangles, but it is not difficult work—just a bunch of baby steps to the final answer.

Examples

1. Here is a simple one:

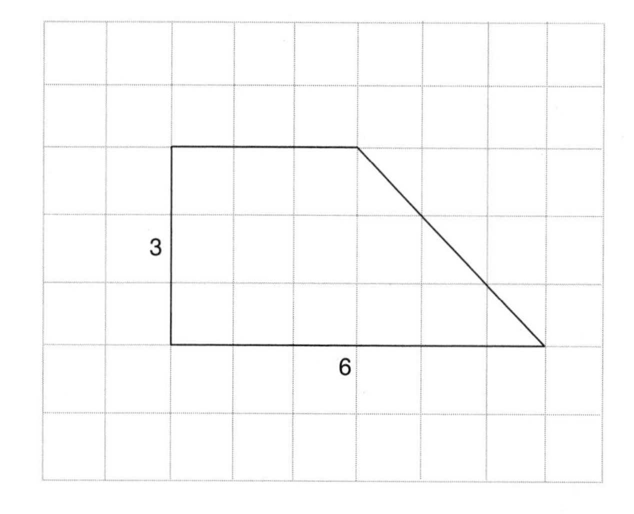

This trapezoid can be cut into a square and a right triangle. Then we find the area of both of those shapes and add them together.

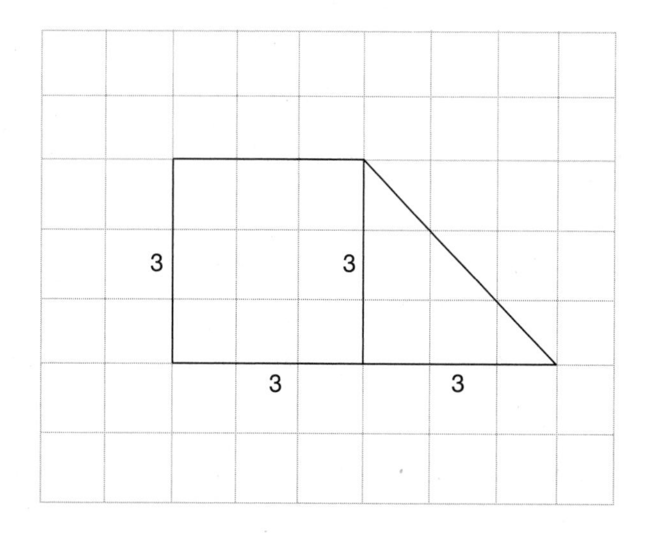

The square is 3 × 3, or 9 square units. The triangle is half of a 3 × 3 square, which is 4.5 square units. So the area of the original figure is 9 + 4.5 = 13.5 square units.

2. The same method can be used for familiar shapes, like this hexagon:

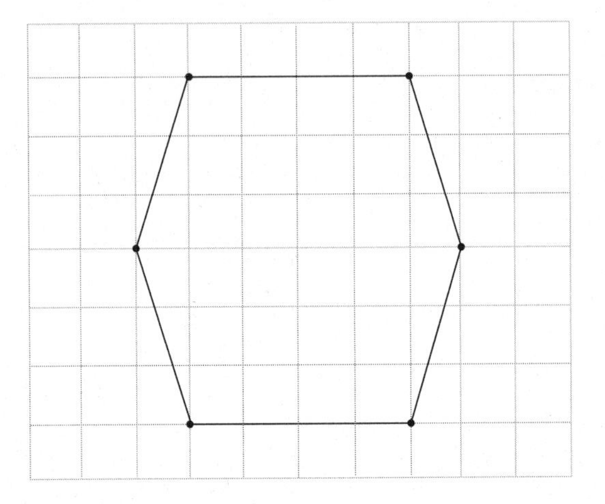

We can cut the hexagon into a rectangle and four right triangles:

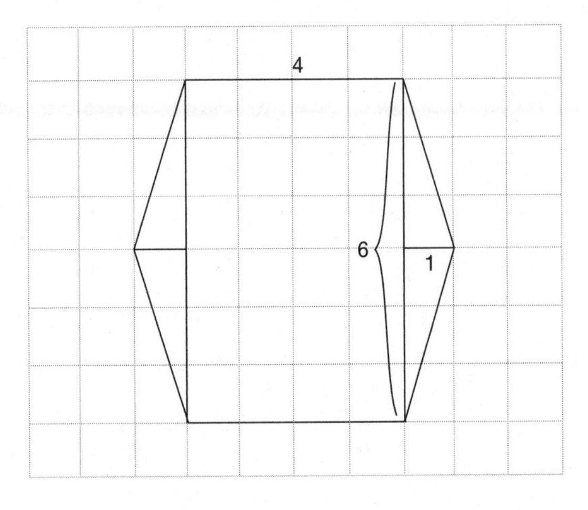

The rectangle is 4 units wide and 6 units long, so its area is 24 square units. The four triangles are all congruent (the same size and shape). They are 3 units tall and 1 unit wide, so the area of each triangle is $\frac{1}{2} \times 1 \times 3 = 1.5$ square units. The total area of the four triangles is 4 × 1.5 square units = 6 square units. Therefore, the area of the entire hexagon is 24 + 6 = 30 square units.

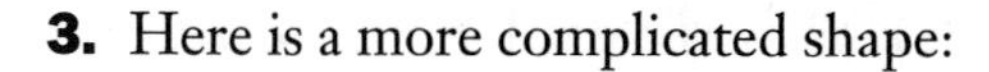

3. Here is a more complicated shape:

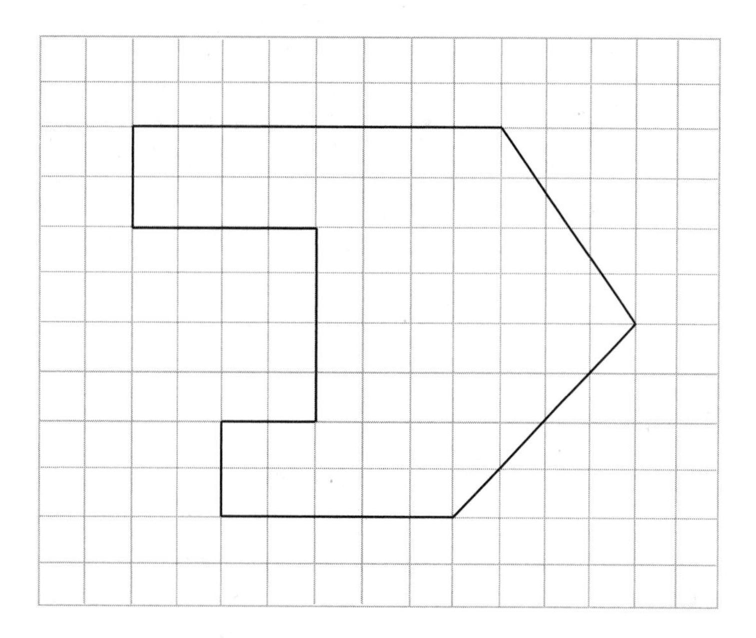

Here is the same shape, divided up, with each area labeled. Since the shape is printed on a grid, it is easy to count or multiply to find the area of each rectangular piece. For the triangles, visualize the triangle doubled into a rectangle and divide by 2.

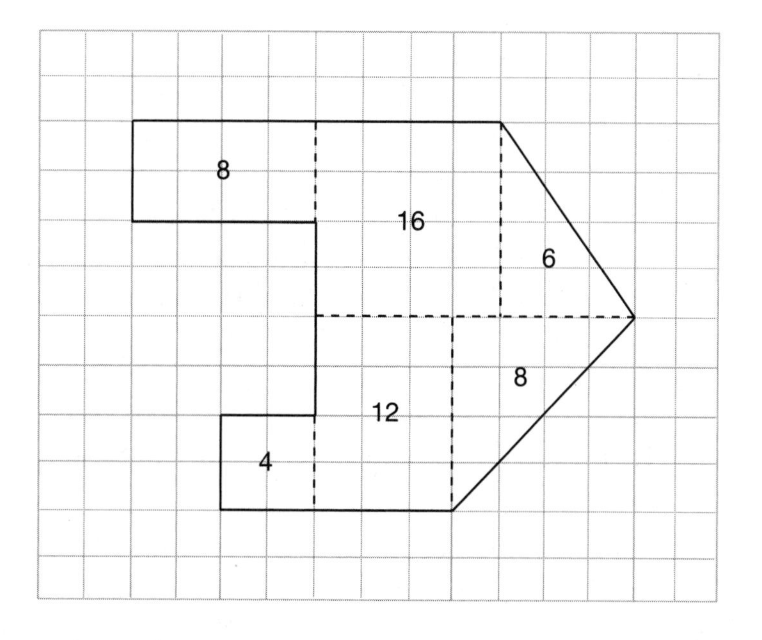

The total area is 8 + 16 + 6 + 4 + 12 + 8 = 54 square units.

TRY MULTIPLE METHODS

Every good ninja knows there's more than one way to slice and dice a polygon. Usually the fewer shapes, the better—but there is no correct way to cut a composite shape up. Find a way that makes sense to you, and do it. A great way to practice with your child is by seeing how many *different* ways you can cut a polygon up, and making sure you get the same total area every time.

Three-Dimensional Measures

Three-dimensional geometry is a great example of a topic that gets far more attention now than it ever did before. Years ago, three-dimensional geometry was seldom covered in middle school. In high school geometry textbooks, it was often the last chapter and therefore the first to get cut out of the course when a class fell behind. Now, middle school students are expected to know a great deal about three-dimensional geometry, including surface area and volume. Fortunately, there are some great visualization techniques that students have access to.

Surface Area

The **surface area** of a three-dimensional object is the total area of the outside surface of the object. It is a two-dimensional measure of a three-dimensional shape, just as perimeter is a one-dimensional measure of a two-dimensional shape. Much like the areas of triangles, trapezoids, parallelograms, and other shapes, surface area is a concept that has often been taught without using visuals. Today's students have the advantage of a more sensible approach, which will help them develop spatial reasoning skills as well as number sense.

The approach that your child will likely see involves *nets*. A net is a two-dimensional representation of a three-dimensional figure. The following image is a net. Each section is labeled and will be used momentarily to find the surface area of a three-dimensional shape.

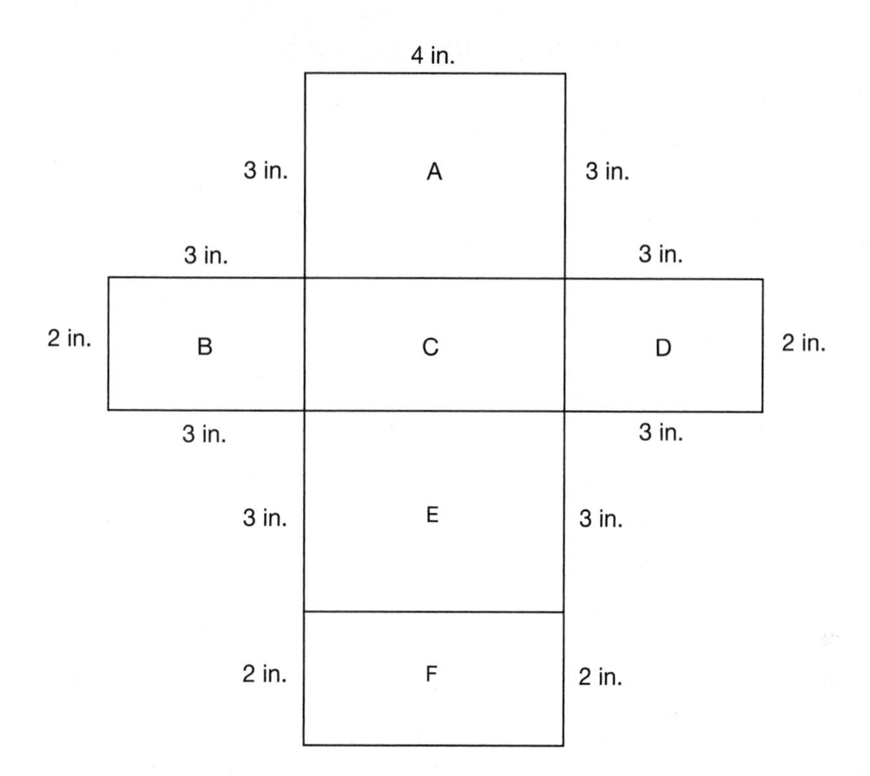

If you trace this shape on a blank paper and cut it out, you can fold along the interior lines, and the paper will take the shape of a rectangular prism (that's a fancy math term that means "box").

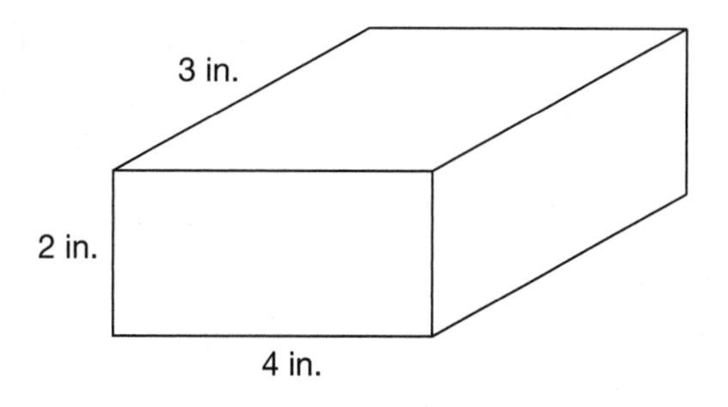

Look back at the original net. The areas marked A and E have become the top and bottom of our box. Their dimensions are 3×4, so each of these has an area of 12 square inches. The areas marked B and D have become the left and right sides of our box, each with an area of $3 \times 2 = 6$ square inches. The front and back are C and F, each with an area of $4 \times 2 = 8$ square inches. The total area of the surface of our rectangular prism is 12 + 12 +

6 + 6 + 8 + 8 = 52 square inches. Of course there are shortcuts for calculating the surface area of a rectangular prism. But visualizing any rectangular prism as a flat net will help your child understand what surface area means and how it is connected to the three-dimensional shape.

Your child will likely get a lot of practice in 6th or 7th grade cutting, folding, and taping nets. The bigger challenge is learning to visualize or draw a net based on an image of a rectangular prism. For example, a word problem might ask your child to figure out how much cardboard is needed to make the cereal box below.

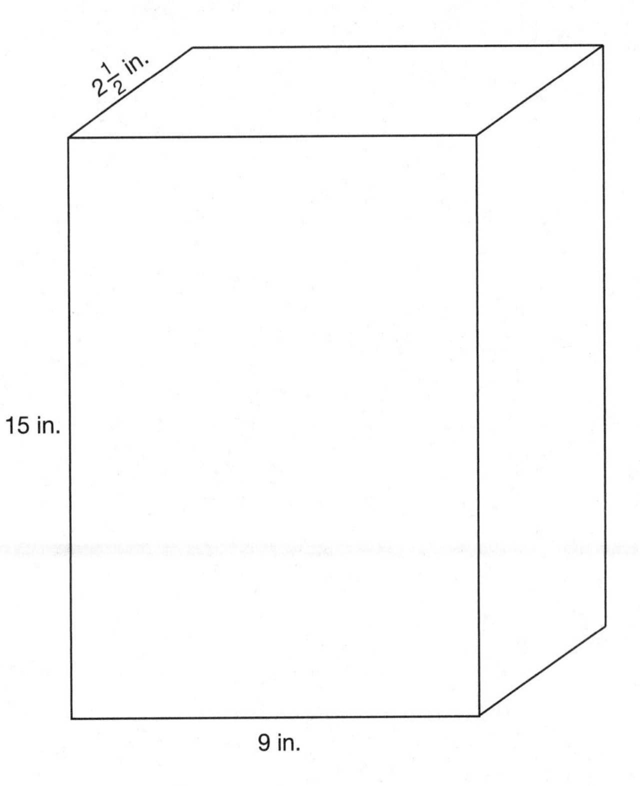

Note that in real life, a box like this would have some overlap—there are flaps inside the box where the different parts are glued together. Usually in math problems, we ignore this fact.

In order to find the surface area, we could imagine cutting the box open (a great thing to actually practice at home) and draw a net for it:

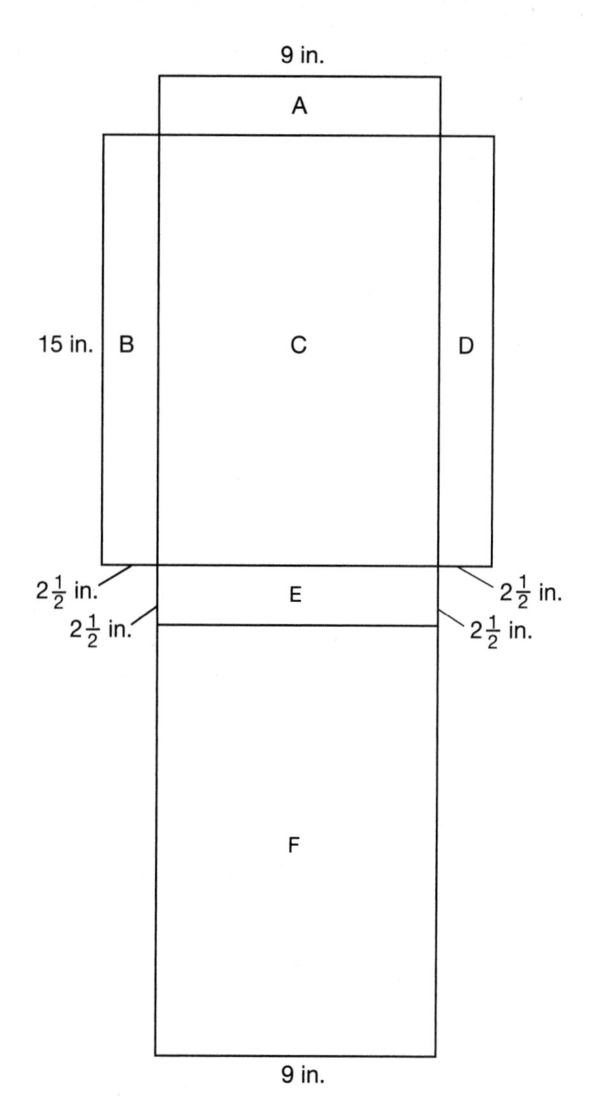

Now we can find the area of each section.

Sections A & E (top and bottom): 9 inches × $2\frac{1}{2}$ inches = $22\frac{1}{2}$ square inches

Sections B & D (left and right sides): 15 inches × $2\frac{1}{2}$ inches = $37\frac{1}{2}$ square inches

Sections C & F (front and back): 15 inches × 9 inches = 135 square inches

Since the sides of the box come in pairs, we can find the total by doubling each area and adding them together:

Surface Area = $2 \times 22\frac{1}{2} + 2 \times 37\frac{1}{2} + 2 \times 135$

Surface Area = $45 + 75 + 270$

Surface Area = 390 square inches.

We could find the total surface area a different way, again using the idea that each face is paired with an identical face on the other side. If we add up the areas of the three different faces and then double the sum, we get the same result:

Surface Area = $2 \times (22\frac{1}{2} + 37\frac{1}{2} + 135)$

Surface Area = 2×195

Surface Area = 390 square inches.

Later, in Chapter 7, we will use the relationship between these two methods as a way of visualizing an important algebra concept, the Distributive Property.

You can also find the surface area of other three-dimensional shapes using a net. Here is a pyramid:

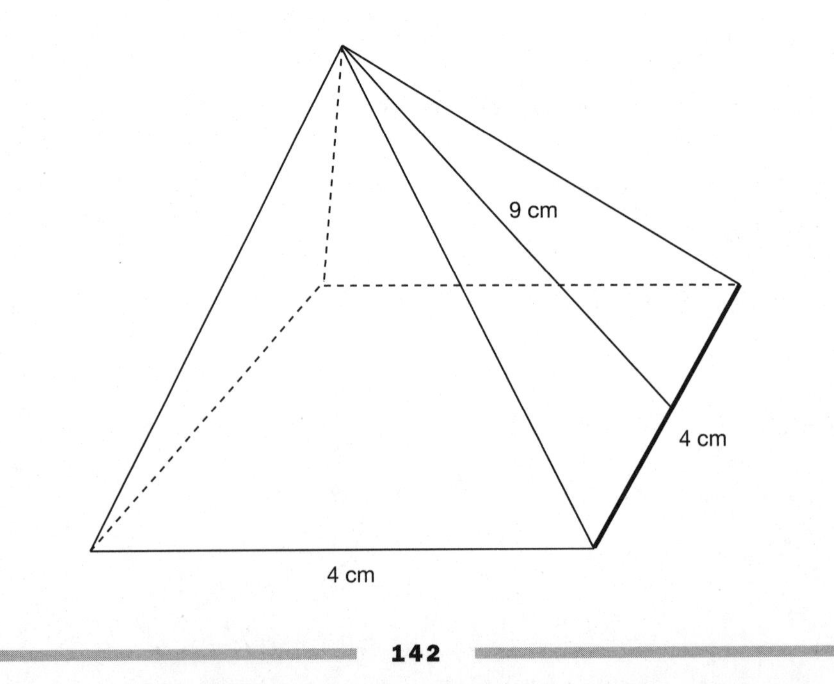

The pyramid has a square base and four congruent triangular faces. The height of the triangle is given, but note that this is *not* the height of the pyramid. In later grades, your child may have to find the height of the triangle based on the height of the pyramid, but not yet. Here is the net for this pyramid:

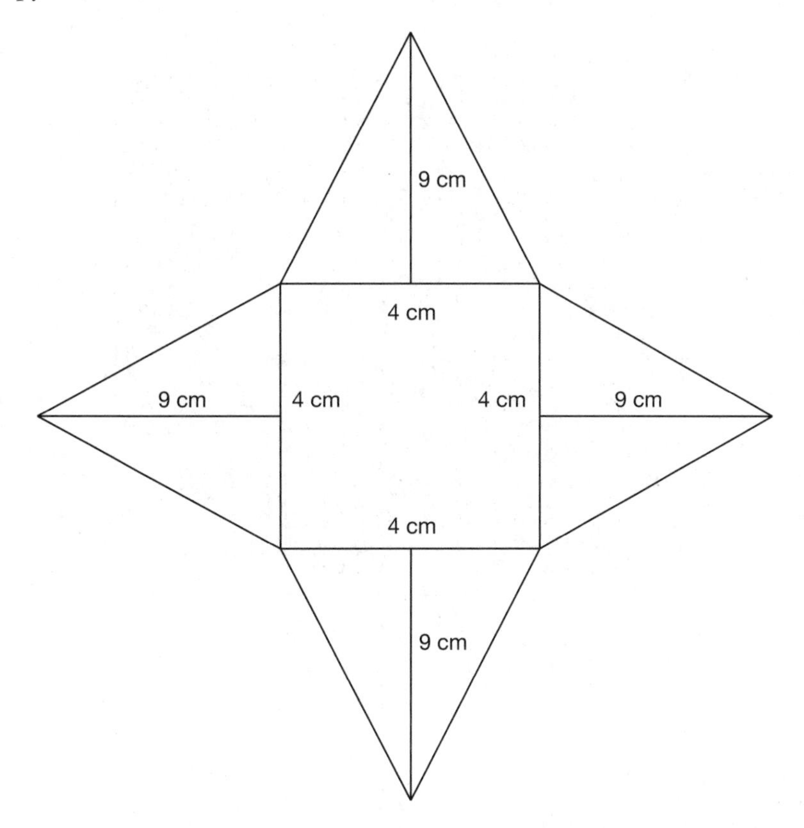

The area of the square base of the pyramid is 4 cm × 4 cm = 16 square centimeters. The area of each triangular face is $\frac{1}{2}$ × 4 cm × 9 cm = 18 square centimeters. There are 4 of these triangles, so the total surface area of the pyramid is

Surface Area = 4 × 18 + 16

Surface Area = 88 square centimeters.

Volume

Whenever we express an area, we use square units because we are measuring how many flat squares of a certain size could cover the area. Similarly, three-dimensional measurements, called **volume**, do not use simple units.

While area is measured in square units, volume is measured in **cubic units**. One cubic inch is the volume of a cube (or box) one inch in length, one inch in height, and one inch in width.

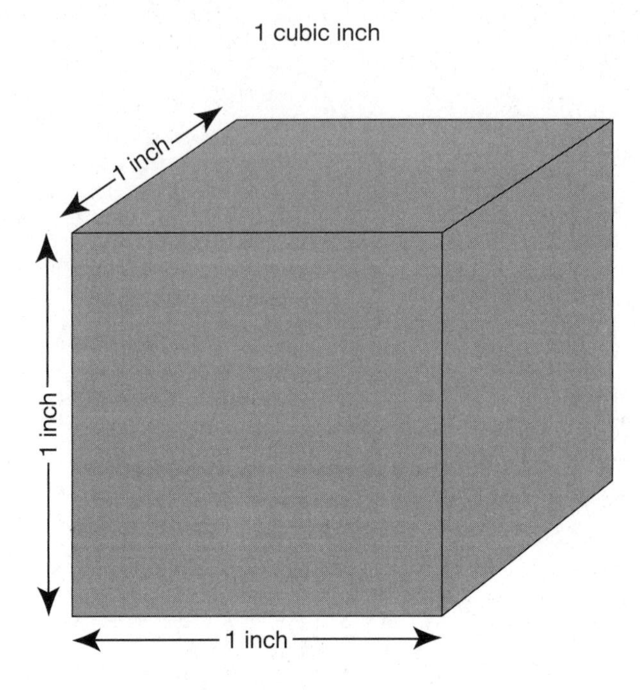

To find the volume of a space, we need to figure out how many cubic units would fit inside the space.

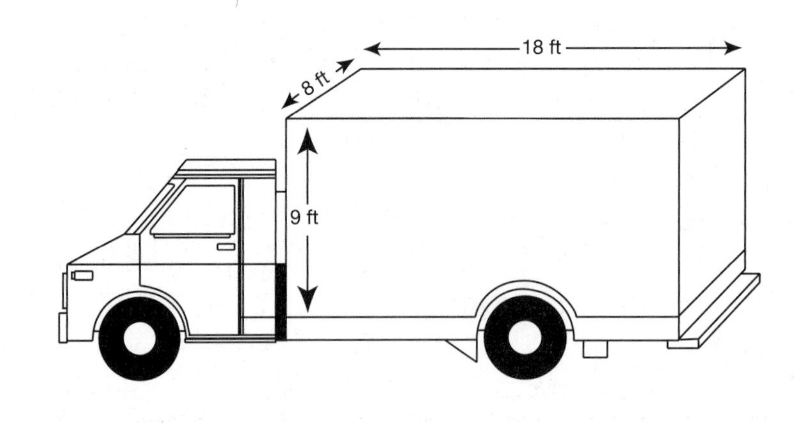

If we were loading this truck full of 1 × 1 × 1 boxes, we could start by loading a layer on the floor of the truck, which is 8 feet wide and 18 feet long. The area of the floor is 8 × 18 = 144 square feet, so we could place 144 boxes on the bottom layer. Since the truck's storage space is 9 feet high, we could put another 8 layers on top of the base layer, each made up of 144 cubic units. The truck could hold a total of 9 × 144 cubes, so the volume of the truck is 1,296 cubic feet.

Of course, the old formula for volume, $V = lwh$, still works. Your child likely learned the approach of visualizing layers in elementary school, and in middle school she will be expected to use this formula to find volumes for rectangular prisms, even if the dimensions are fractions.

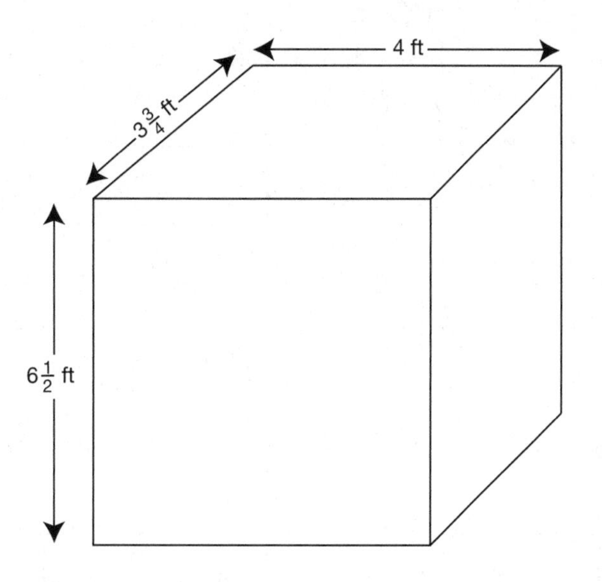

Here we have a rectangular prism with a height of $6\frac{1}{2}$ feet, a length of $3\frac{3}{4}$ feet, and a width of 4 feet. To find the volume, we need to multiply:

$$V = 6\frac{1}{2} \times 3\frac{3}{4} \times 4$$

If your child's fraction sense is strong, then she will likely see that it would be easy to multiply $3\frac{3}{4}$ by 4 first, eliminating one of the fractions. $3\frac{3}{4} \times 4$ can be done easily by breaking up the $3\frac{3}{4}$. $3 \times 4 = 12$, and $\frac{3}{4} \times 4 = 3$, so $3\frac{3}{4} \times 4 = 12 + 3 = 15$. Now we multiply $6\frac{1}{2}$ by 15 in the same way. $6 \times 15 = 90$, and $\frac{1}{2} \times 15 = 7\frac{1}{2}$, so $6\frac{1}{2} \times 15 = 97\frac{1}{2}$. The volume of this rectangular prism is $97\frac{1}{2}$ cubic feet.

Scale Drawings and Similarity

Scale drawings are one of the most commonly used topics in geometry. Maps are an example you'll recognize—whenever you use your GPS to get someplace new, you are using a scale drawing. If you and your child have ever built a model, the instructions were probably a scale drawing and the model itself would have been to scale with the real thing. Scale drawings are also used in many professions—architects, engineers, interior designers, inventors, machinists, and many others use blueprints, schematics, and other kinds of scale drawings to design, plan, and understand the projects they are working on.

A **scale drawing** is a diagram that represents something else that is a different size. Distances between objects in a scale drawing are all proportional to the real-life distances between those same objects.

The following map has a scale given as "1 inch = 3 miles." This means that if two locations are 1 inch apart on the map, then they are 3 miles apart in real life.

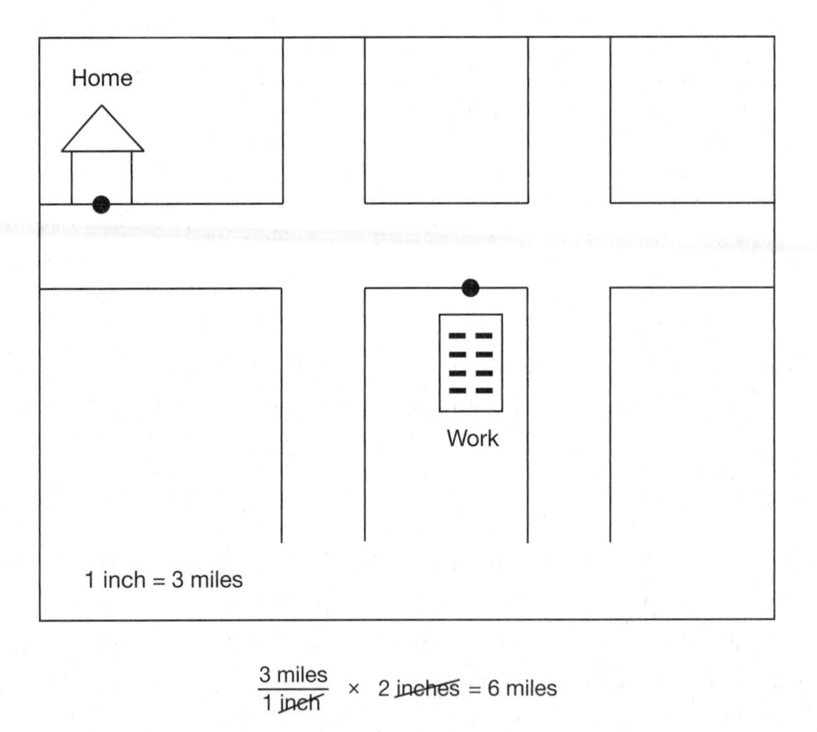

The distance from home to work on the map above measures 2 inches, which means the real-life distance between these locations is $3 \times 2 = 6$ miles.

We can use a ratio table to find other real-life distances based on map distances:

Map Distance	1 inch	2 inches	3 inches	10 inches	15 inches	$3\frac{1}{4}$ inches
Real-Life Distance	3 miles	6 miles	9 miles	30 miles	45 miles	$9\frac{3}{4}$ miles

In Chapter 4, we learned about the constant of proportionality. Another name for this, which is used in geometry, is **scale factor**. The scale factor of a scale drawing is the ratio by which you would multiply distances to convert from the map to real life, or vice versa. In our first example, since every inch represents 3 miles, our scale factor is $\frac{3 \text{ miles}}{\text{inch}}$, or 3 miles per inch.

A NOTE ABOUT UNITS

We have written the scale factor as $\frac{3 \text{ miles}}{\text{inch}}$, but it might be easier for your child to think of it as just 3, and change the units at the end. For our purposes, that is fine, but be aware that in high school, your child's math and science teachers will likely be much stricter about units.

If we simply multiply by 3, what we are technically saying is that everything in real life is just 3 times larger than it is on the map. Obviously this is not true, and if it were, our map probably would not be very useful, because every mile in real life would be $\frac{1}{3}$ of a mile on paper (that's about 1,700 feet). When we use units, *3 miles per inch*, we are not only multiplying the number of inches by 3—we are also turning inches into miles, which is like multiplying by 63,360, the number of inches in a mile. Since each inch is being converted to 3 miles, we are really multiplying by a factor of 3 times 63,360, which is 190,080. Distances in real life are almost 200,000 times the distances on the map!

Here's another example.

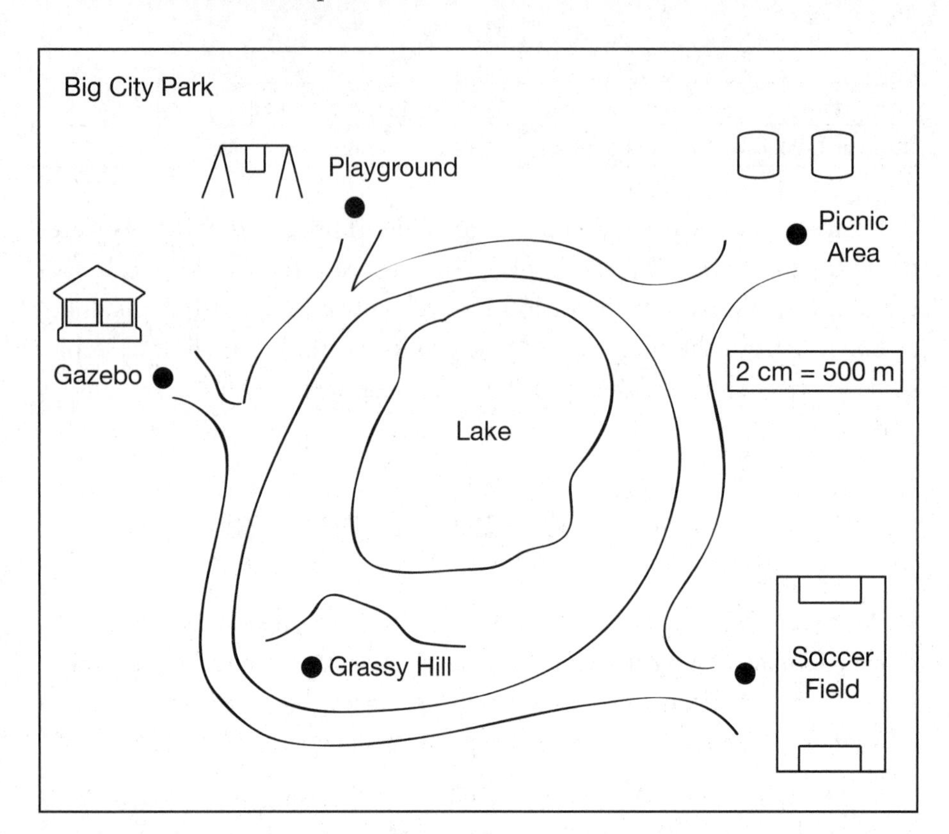

In this map, every 2 centimeters represents 500 meters. If the picnic area is 6 cm away from the playground, how far is it in real life?
In this case, we need to use the ratio $\frac{500\text{ m}}{2\text{ cm}}$, which is our scale factor. We can simplify this to a scale factor of 250 meters per 1 centimeter.

$$250 \times 6 = 1{,}500$$

So the distance from the picnic area to the playground is 1,500 meters.

> **"NOT DRAWN TO SCALE"**
>
> **Not every map or drawing is a scale drawing. Calling something a scale drawing means that its distances are all in the same proportion to the real-life place or object. You will often see diagrams labeled "not drawn to scale" when this is not the case, as in the area models we used for multiplication earlier in this chapter.**

Similar Figures

In middle school, your child will work with scale drawings a lot. While maps and other practical scale drawings are very useful in real life, simpler and more abstract scale drawings, called *similar figures*, are very helpful as a representation for other topics in math.

Similar is one of those funny words that means one thing in everyday life and has a different meaning when used in a mathematical context. In geometry, we say that two figures are **similar** if they have exactly the same shape and proportions, but are different sizes.

For example, the two rectangles below are similar:

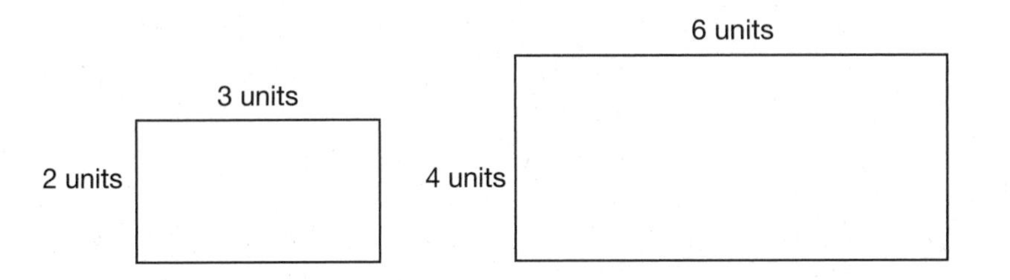

The width of the smaller rectangle (3 units) is half the width of the larger rectangle (6 units). The length of the smaller rectangle (2 units) is half the length of the larger rectangle (4 units). In both rectangles, the width is 1.5 times larger than the length: $2 \times 1.5 = 3$, and $4 \times 1.5 = 6$. These are the features of similar figures.

Similar Figures as a Visualization Tool

Some math teachers insist that their students master the concept of proportionality that we saw in Chapter 4 before they can move on to similarity. Understanding how proportions work and knowing how to compute with proportions makes learning similarity a breeze for some students. But there are other students—those visual learners—who do not properly understand proportions *until* they learn about similarity. The visual nature of similar shapes helps the concepts of proportionality click for these students.

Suppose you need to solve the proportion $\frac{3}{8} = \frac{9}{x}$. We learned several methods for doing this in Chapter 4. Now, we can use similar rectangles to solve it. We will use 3 and 8 as the dimensions of one rectangle, and 9 and x as the corresponding dimensions of a similar rectangle:

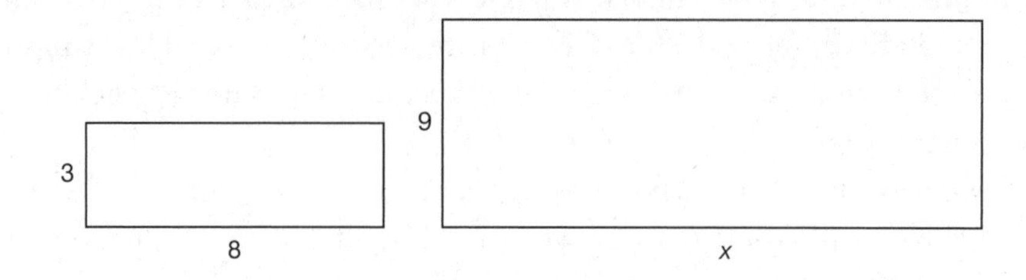

Since 3 × 3 = 9, 3 must be our scale factor for these two rectangles.

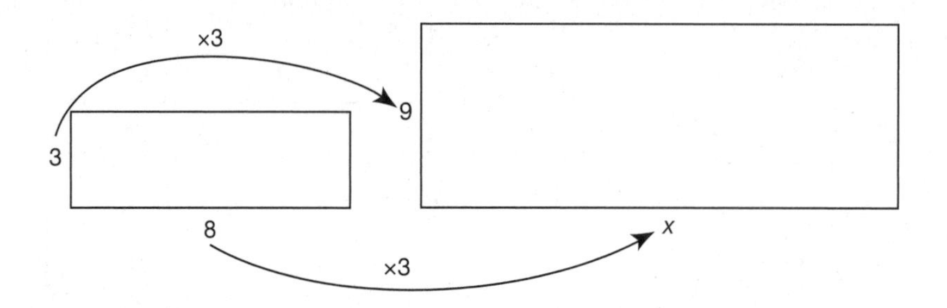

$x = 8 \times 3 = 24$. If your child struggles with the proportion concepts presented in Chapter 4, a simple visual example can make a big difference in her understanding. There is nothing wrong with relying on this method to solve proportion problems.

Here's a more difficult example. Let's solve $\frac{154}{66} = \frac{x}{12}$. Note that in this example, as in the last one, the drawing will not be to scale. If we tried to draw one rectangle with a length of 12, and another with a width of 154, either the first rectangle would be extremely small, or the second would be too large for the page. As much as we would like these diagrams to have the correct proportions, sometimes it is not practical. As long as everything is labeled properly, it does not matter.

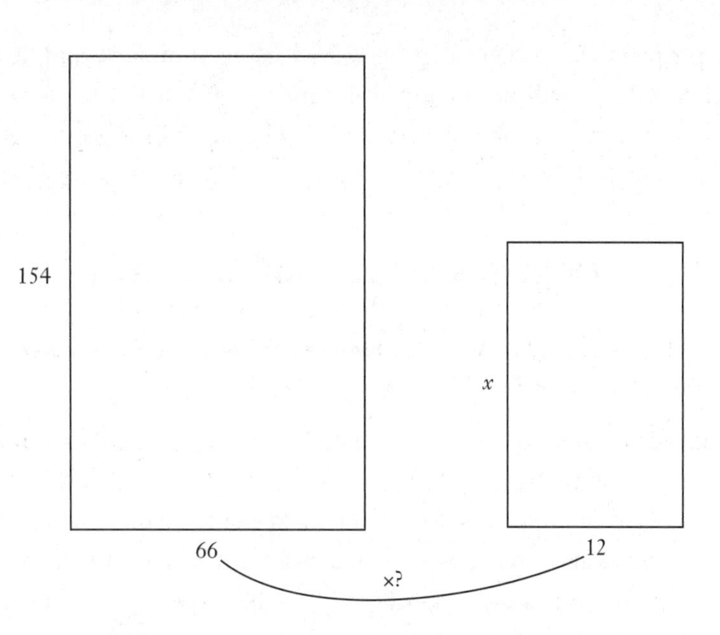

In this case, it's not easy to see what the scale factor is because we don't know a number that can multiply by 12 to make 66. We can divide to find the scale factor: $66 \div 12 = 5.5$. The scale factor is 5.5, so x needs to be multiplied by 5.5 to get 154. But how can we multiply x by 5.5 if we don't know what x is? We work backwards:

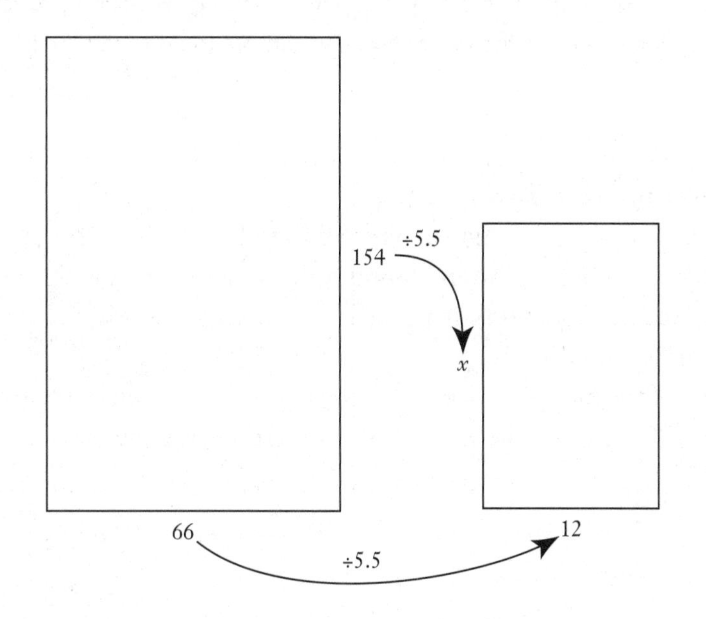

$154 \div 5.5 = 28$, so $x = 28$.

Any proportion problem can be solved using similar rectangles. If you think this method will help your child understand proportions better, try applying it to any proportions problem—whether it is a homework problem or one of those everyday-life proportions we discussed in Chapter 4.

MORE STRATEGIES FOR SUCCESS

If your child is struggling with this topic (as many students do), there are a few strategies you can try to make it easier for her.

One idea is to take problems with complicated, "ugly" fractions and relate them to simpler problems. "What if these were whole numbers?" Make up some "easier" numbers and see if your child can solve the proportion that way. Then walk your child through the same process, but with the more difficult fractions. In this way, she will see that the process is the same, even if the numbers are more difficult to handle. If you find that it's the fractions that are giving your child trouble rather than the concept of proportions, then it is a great time to practice more with fractions (with no pressure and lots of rewards), since mastery of fractions will be essential throughout the rest of your child's math career (not to mention life).

The most important thing when your child is struggling is never to let her give up. Persistence is the key to success in math (not to mention life).

Similar Figures and Area

When studying similar figures, your child will be asked to change the scale of an original figure and then compute the area of the new figure (called the **image**). Many students struggle conceptually when dealing with area in scale drawings. If the scale of a drawing is 3:1, then the dimensions of every part of the picture are scaled up by a factor of 3. But, as the following example will illustrate, the area does not increase by a factor of 3.

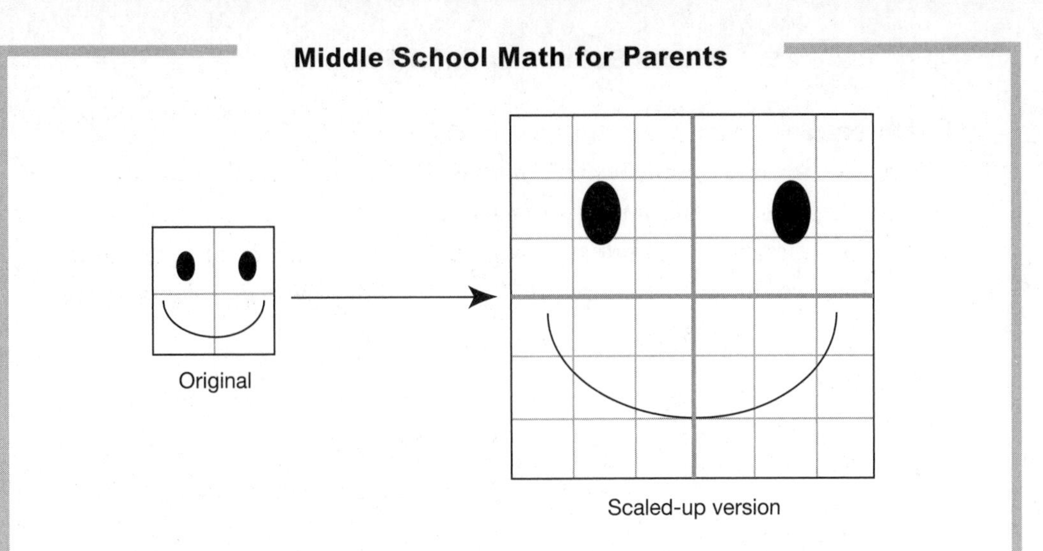
Original

Scaled-up version

In the original figure, the face is a 2 × 2 square, with an area of 4 square units. Using a scale factor of 3, the scaled-up version is 6 units long on each side. The length and width both tripled. But look at the area: The scaled-up figure has 36 square units in it. The area multiplied by 9.

Notice that if you cut the scaled-up figure into 4 equal sections, each section is 9 square units. These 4 sections correspond to the square units in the original figure. Since each dimension was multiplied by 3, every 1 square unit in the original has grown to 9 square units in the image.

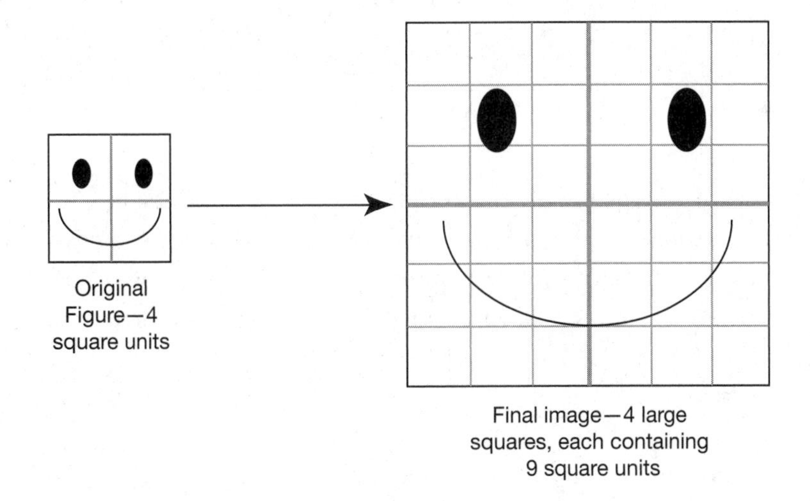
Original Figure—4 square units

Final image—4 large squares, each containing 9 square units

When scaling a figure, the area of the figure changes by a factor of the scale factor multiplied by itself.

Since every dimension was multiplied by a scale factor of 3, the area increased by a factor of 3 × 3, which is 9.

Example

A rectangle measures 2 inches by 6 inches. What would be the area of the new rectangle after scaling up by a factor of 5?

Since the original rectangle is being scaled up by a factor of 5, the new dimensions would be 10 inches and 30 inches. The new area is 10 in. × 30 in. = 300 square inches.

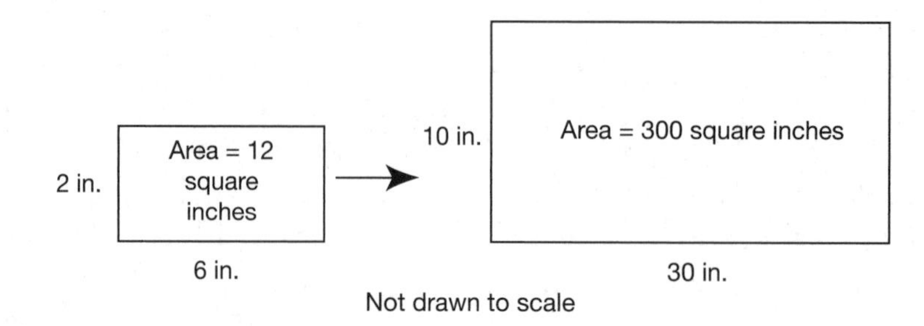

Notice that the original rectangle had an area of 12 square inches. Since the scale factor is 5, each unit square becomes a 5-by-5 square, or 25 square units. Therefore, the area increases by a factor of 25:

25 × 12 square inches = 300 square inches

Visualizing Word Problems

Visualization is an essential tool for helping students understand new math concepts. It can also be a great strategy for solving word problems, which often give students *and* parents trouble. Let's look at a few word problems from a variety of topics to see how visualization can help solve them.

Examples

1. Alex is watching a superhero movie on a smartphone screen. In one scene, Ninja Dude is 1.5 inches tall onscreen, and his partner, Polly Gone, is 1.25 inches tall (if you can even see her—she's usually invisible). On a movie theater screen, Ninja Dude is 18 feet tall. How tall is Polly Gone on the theater screen?

This one is simple to visualize because it is a scale drawing problem. Begin by drawing the picture:

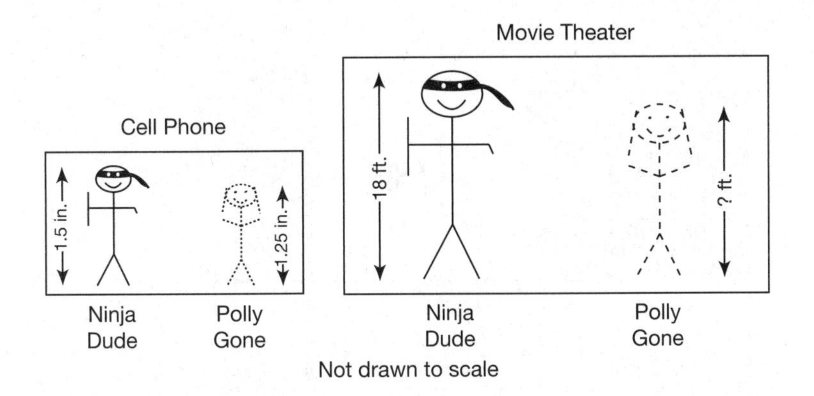

Now it's a simple matter of finding the scale factor. We can find the scale factor by dividing Ninja Dude's heights: $18 \div 1.5 = 12$, so the scale factor is 12 feet per inch—every inch on the phone screen is scaled up to 12 feet on the movie theater screen. To find Polly Gone's height, just multiply: $1.25 \times 12 = 15$. Polly will be 15 feet tall.

2. Allie is baking mini-pies for her math class's Pi Day party. The recipe for 4 dozen mini-pies requires $1\frac{1}{2}$ cups of sugar. If Allie wants to make 1 dozen mini-pies only, how much sugar should she use?

Let's draw a picture!

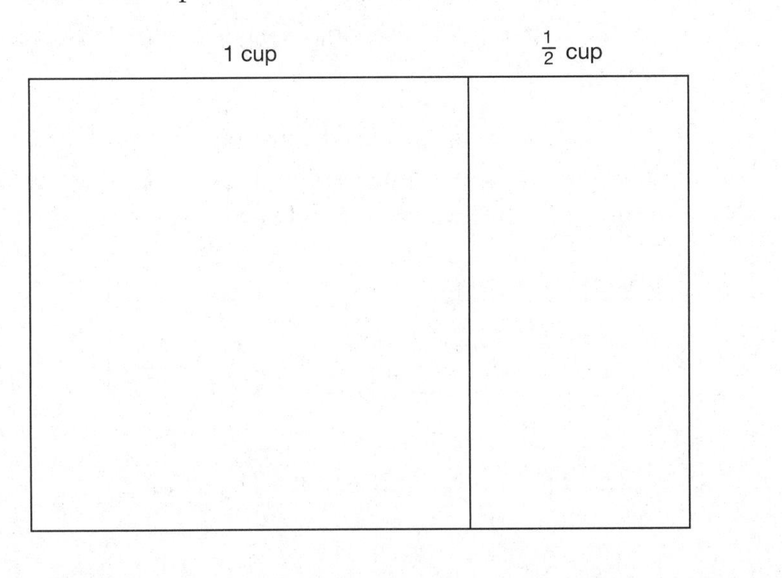

We have shown the $1\frac{1}{2}$ cups of sugar needed for 4 dozen cookies. Now let's cut each section into 4 equal pieces to represent the amount that will be needed for 1 dozen mini-pies.

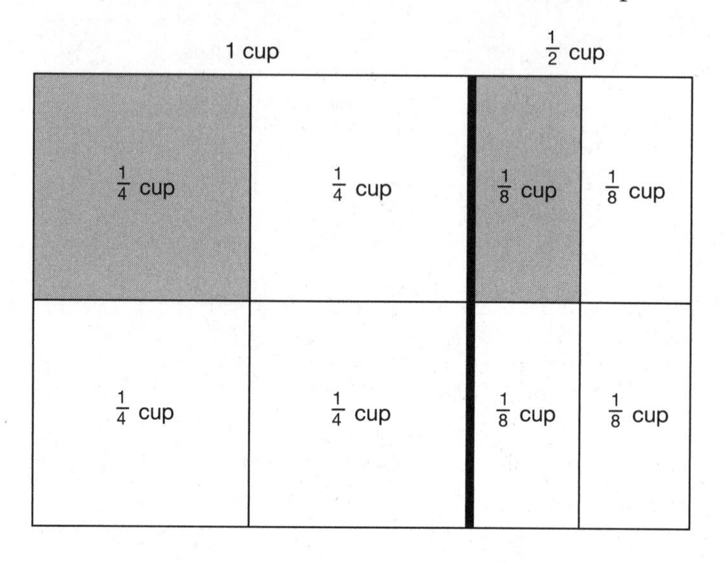

The shaded areas represent the sugar Allie will need—$\frac{1}{4}$ of a whole cup and $\frac{1}{8}$ of half a cup. The larger shaded area is 1 cup ÷ 4 = $\frac{1}{4}$ cup, and the smaller shaded area is $\frac{1}{2} \div 4 = \frac{1}{8}$ cup. So altogether, Allie will need $\frac{1}{4} + \frac{1}{8} = \frac{3}{8}$ cups of sugar.

3. Allen lives 0.6 miles from school. One day he was walking to school, and when he was halfway there he realized he didn't have his homework. So he turned around to go home. When he got halfway home from the point where he turned around, he found his homework on the sidewalk, so he picked it up and walked the rest of the way straight to school. What is the total distance Allen walked?

For this one, we definitely need a diagram. Let's start by drawing the distance from home to school. Then, we'll use arrows to show how far Allen walked on each leg of his journey.

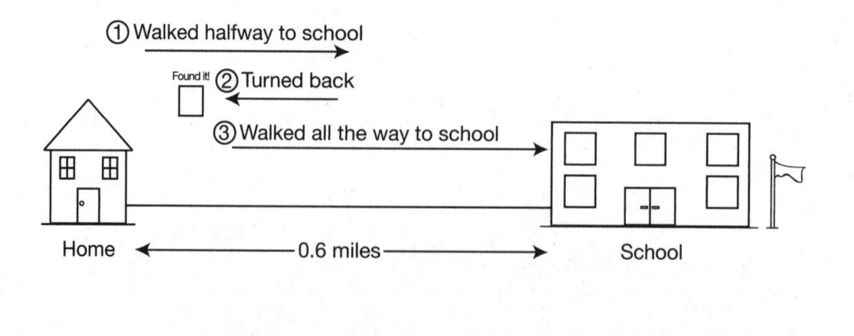

In this diagram, each arrow is labeled and numbered. When making complicated diagrams, labeling is critical. All we need to do now is find the length of each arrow, and add them up.

Arrow #1 is half of 0.6 miles, which is 0.3 miles.

Arrow #2 is half of half the length, so it is $\frac{1}{4}$ of 0.6 miles, or 0.15 miles.

Arrow #3 is $\frac{3}{4}$ of 0.6 miles, or 0.45 miles.

Allen's total distance is $0.3 + 0.15 + 0.45 = 0.9$ miles.

There are many other ways to solve this problem, but this is one method that shows you how a visual diagram can be used.

Summary

In this chapter, we have covered some very important topics in geometry—length and area, volume and surface area, and scale drawings. We have seen that simple geometry concepts can be used as representations for a variety of problems in other areas of mathematics, including numbers and operations, and algebra. While some students are identified as "visual learners," all students can benefit from seeing a problem or a new idea represented visually, even if that concept does not seem to have anything to do with geometry. Much of the most important mathematics that we know would never have been discovered without connections to visual concepts like the ones we used in this chapter. When your child is struggling with any new concept from now on, she can always fall back on a visual representation for it. If her teacher does not provide a visual, then she should ask. If that fails, there's always the Internet.

We have also solved some word problems on a few different topics (fractions and decimals) that might not seem visual at first glance, but which become much easier to solve once you use a diagram. Drawing a picture for every word problem is a good habit to get your child into as soon as possible.

In Chapter 6, we continue our work with geometry, covering a variety of topics including angles, the Pythagorean theorem, transformations, and circles.

The Joy of Geometry

Having Fun with Math and Why Fun Matters

Believe it or not, many people consider math fun. Some people like seeing the way numbers fit together; others enjoy discovering geometric patterns. Some people see an algebra problem as a puzzle to solve; others find pleasure in connecting mathematical ideas to everyday life.

If you think about the activities you and your child enjoy doing together, you may find that even if math doesn't make the list, some of your favorite pastimes have some mathematical component to them. Soccer, basketball, and other sports are all about geometry. Music is made of patterns. Games are full of strategic thinking, pattern recognition, logic, and probability. If you like crafts, you probably use geometric patterns all the time. You use estimation skills and lots of other mental math whenever you take your child fishing, bowling, shopping, or on a road trip.

Math finds its way into most of the things we do, and while math may not be the reason we enjoy our favorite activities, it is an unavoidable part of them.

The Importance of Having Fun

If you were to make a list of things you are good at and a list of things you enjoy doing, there's a good chance there would be a lot of overlap between the two lists. In Chapter 1, we saw the cyclical relationship between attitude, practice, and performance. Similarly, enjoyment, practice, and success in any endeavor feed off each other.

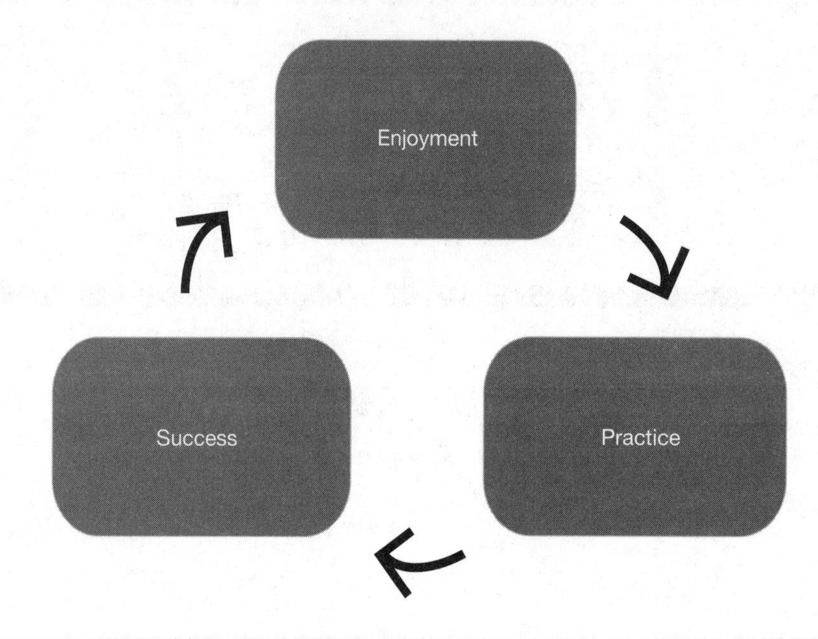

Suppose your child is a musician. He plays his instrument all the time, constantly practicing and getting better at it. As he gets better, he feels more and more successful, which increases his enjoyment of music. This progression is exactly what we would like to see happen with math. The problem is that every time he is faced with math homework, he's more likely to pick up his guitar instead. He hasn't been as successful at math (yet), so he doesn't get as much enjoyment out of it . . . so he avoids it.

The cycle has to start somewhere. If you want your child to be successful in math, he must practice. If you want him to practice, you could find ways to force him, but that won't work as well as helping him enjoy it. But enjoyment comes from success, right?

It often happens that students who don't like math in general find geometry more interesting than other math topics. As we discussed in Chapter 5, all students are visual learners to some extent. If your child has not been taught math in a visual way, he may have become discouraged by so many abstract, meaningless ideas. If your child does not like math, then geometry can be a great starting point for the enjoyment of mathematics because it can be an opportunity to learn visually and connect ideas to something concrete. In this chapter, we will focus on a few of the most engaging topics in the middle school curriculum. These are topics that help many students build confidence, and they are a great opportunity to help your child see the relationship between enjoyment, hard work, and success.

Angle Relationships

Many 7th and 8th grade students enjoy learning about angle relationships. Because the problems are so visual, some students are able to solve the problems intuitively. Then, as the problems build up in difficulty, some students who have not previously enjoyed math begin to look forward to the challenge of finding simple relationships in complex diagrams.

Review of Basic Angles

Let's do a quick review of some angles your child most likely learned in elementary school.

An **acute angle** is an angle that measures more than 0° and less than 90°.

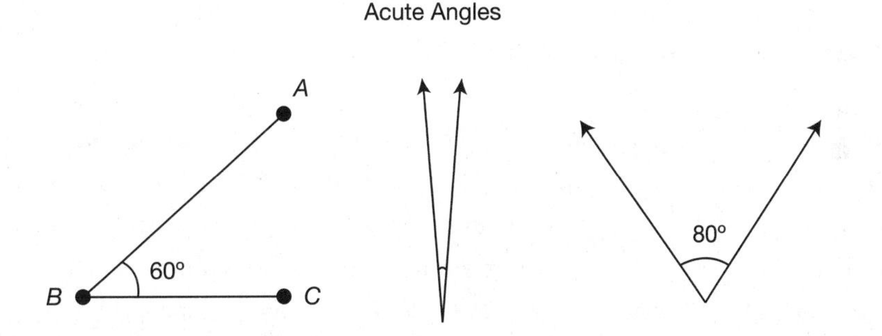

A **right angle** is an angle that measures exactly 90°. Right angles are often labeled with a square in the corner.

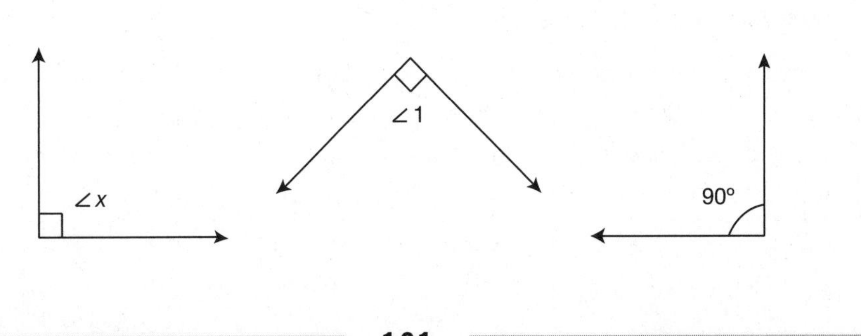

An **obtuse angle** is an angle that measures more than 90° and less than 180°.

Obtuse Angles

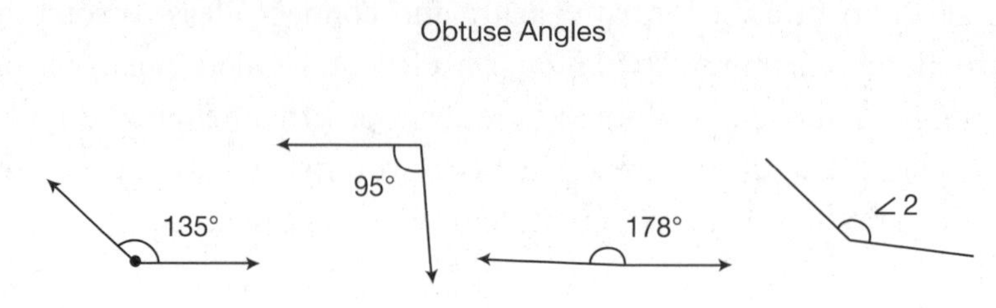

And finally, a **straight angle** is an angle that measures exactly 180°. This one often confuses students because it is a straight line and doesn't look like an angle, but recognizing it is very useful for the work we are about to do.

Straight Angles

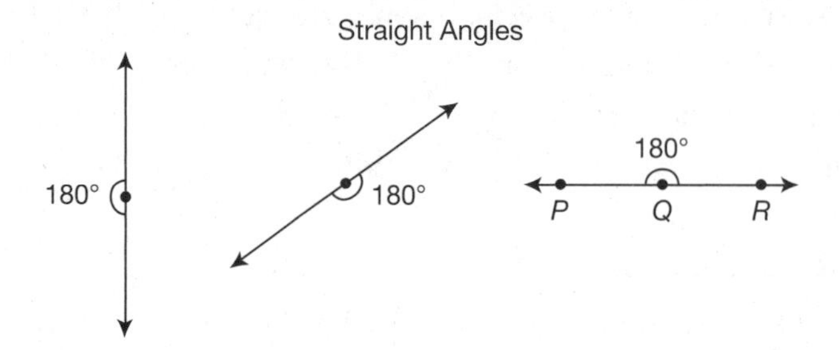

NAMING ANGLES AND ANGLE MEASURES

Many students struggle with naming angles. In the previous examples, several different naming conventions are used. Some angles are named by a single letter, like $\angle x$, which is a right angle. Others are named by a number, like $\angle 1$, which is a right angle, and $\angle 2$, which is obtuse. Sometimes angles have no name but have their measures labeled, like the 95° obtuse angle. Finally, there are two angles that are named by *three* points. $\angle ABC$ is an acute angle. $\angle PQR$ is a straight angle. In both of these angles, the middle letter in the name is the vertex, the point where the two lines (or rays or line segments) of the angle meet.

As you are working with your child on angles, ask him to name each angle. If he has to write an explanation for his answers, make sure his explanations refer to angles by name.

When we want to write the number of degrees in an angle like $\angle 1$, we write $m\angle 1 = 90°$. The letter *m* stands for the *measure* of an angle, which is how many degrees are in the angle. Many students mix up the notation $m\angle 1$ with $\angle 1$. $\angle 1$ is the name of the angle, and $m\angle 1$ is its measure.

Complementary, Supplementary, and Vertical Angles

Complementary angles are two angles that can fit together to form a right angle. They may be adjacent (side by side), in which case you can easily see the right angle. Sometimes they will be separated, in which case you cannot see the right angle, but their measures will add up to 90°.

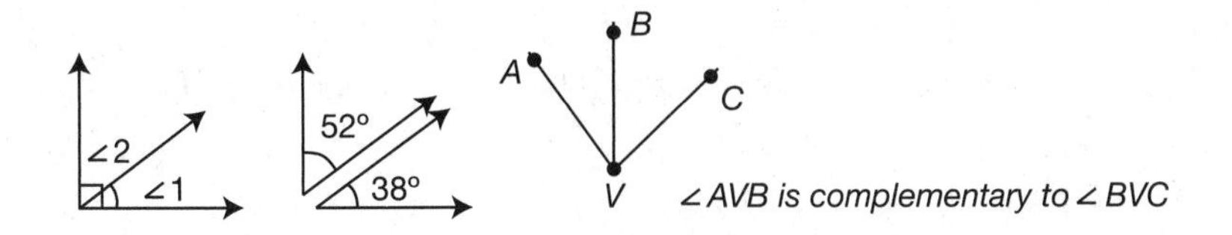

Supplementary angles are two angles that can fit together to form a straight angle. They may be adjacent, in which case we call them a **linear pair** because they form a line. If they are separated, their measures will still add up to 180°.

Examples of Supplementary Angles

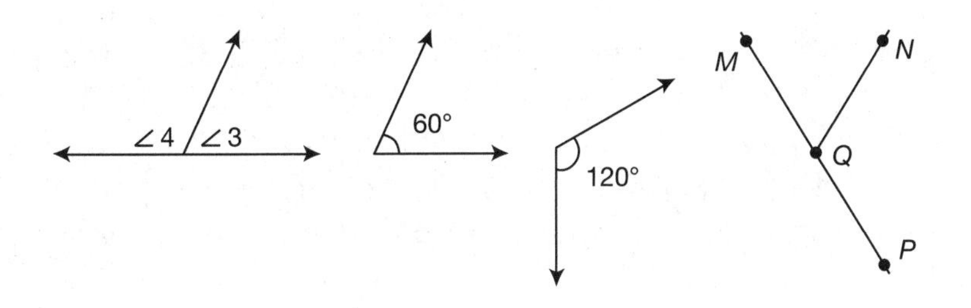

Vertical angles are *two* angles that are formed when two lines intersect. Lines go on forever in both directions, so whenever two lines cross, they form a large "X." Vertical angles are two angles that face opposite directions. Note that the term *vertical angles* has nothing to do with *vertical lines*, which are lines that point straight up and down. In this case, the word *vertical* refers to the fact that the two angles share their vertex.

Examples of Vertical Angles

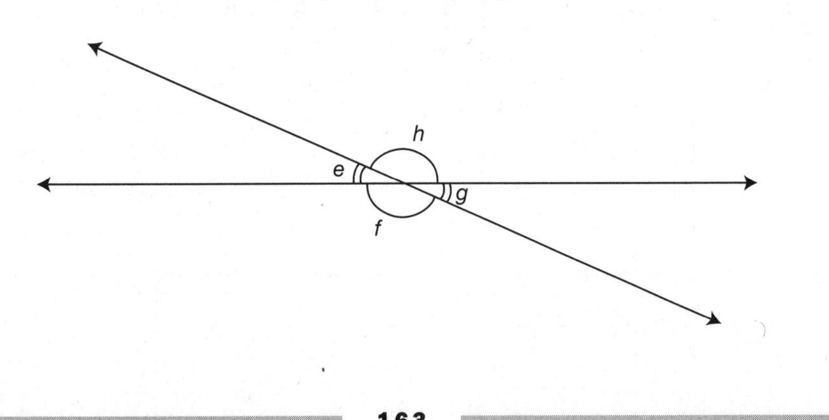

In this diagram, there are four angles (there will always be four angles when two lines cross). There are two pairs of vertical angles: e and g form one pair and h and f form the other. Notice that each angle has the same measure as its vertical angle: If $m\angle e = 35°$, then $m\angle g = 35°$.

Also notice that angles e and h form a linear pair, so they are supplementary. Therefore $m\angle h = 180° - 35° = 145°$. There are four pairs of supplementary angles altogether in this diagram—can you find them all?

One misconception many students have about examples like the previous one is that all four angles are considered a single set of vertical angles. They think vertical angles follow the same pattern as complementary and supplementary angles, and that vertical angles must add up to 360°. Check to see that your child understands that vertical angles are only the *two* angles that point in opposite directions, and that they have equal measures. They can add up to anything between 0° and 360°.

SPEAKING THE LANGUAGE

When describing any angle relationship, we say how one is related to the other. In the previous examples, $\angle AVB$ is *complementary* to $\angle BVC$. The 60° angle is *supplementary* to the 120° angle. $\angle g$ is *vertical* to $\angle e$. You could also say that angles g and e are vertical to each other. Angle relationships are a great topic for teaching students the importance of precise mathematical language.

Reasoning with Complementary, Supplementary, and Vertical Angles

Your child will be given complicated diagrams and asked to find the measures of some angles, given the measures of other angles. In order to do this, one must be able to recognize complementary, supplementary, and vertical angles, and use them to figure out other angle measures.

Examples

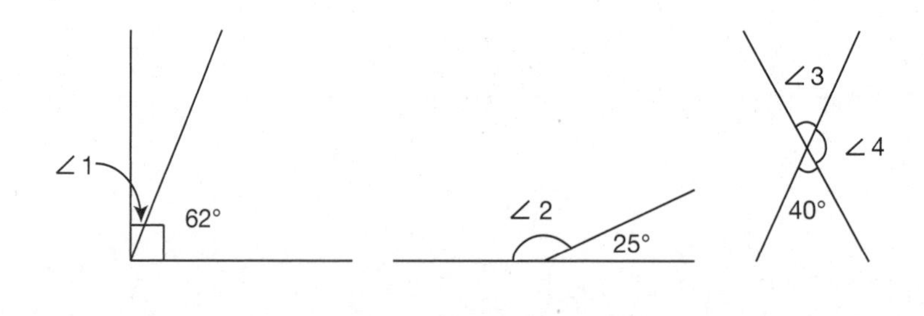

1. The measure of angle 1 ($m\angle 1$) is 28°, since $\angle 1$ is complementary to 62°: $90° - 62° = 28°$.

2. $m\angle 2 = 180° - 25° = 155°$, since $\angle 2$ is supplementary to 25°.

3. $m\angle 3 = 40°$, since $\angle 3$ is vertical to 40°.

4. $m\angle 4 = 140°$, since $\angle 4$ is supplementary to $\angle 3$, which measures 40°.

   Your child might find $m\angle 4$ in a different way. Some students reason that since $\angle 3$ is vertical to 40°, then it must have the same measure. Also, $\angle 4$ and the unlabeled angle are vertical to each other, so they must have the same measure. Since all four angles add up to 360°, you can subtract 80° (the sum of the two 40° angles) from 360°, and divide by 2 to find the $m\angle 4$:

   $m\angle 4 = (360° - 80°) \div 2$
   $m\angle 4 = 280° \div 2$
   $m\angle 4 = 140°$

As a teacher, I look forward to the moment every year when one of my students comes up with this method on his own, because my students become energized as they try to make connections between the two methods. They ask great questions, like "Will it work every time?" "Why does that work?" and "Is that the same as subtracting from 180°?" These are exactly the kinds of questions mathematicians should be asking, and they usually come from the students I least expect. At this moment, there are typically a handful of students who are having fun in math for the first time in a long time.

A More Challenging Example

So far we have used complementary, supplementary, and vertical angles in isolation. Now let's try a more interesting diagram.

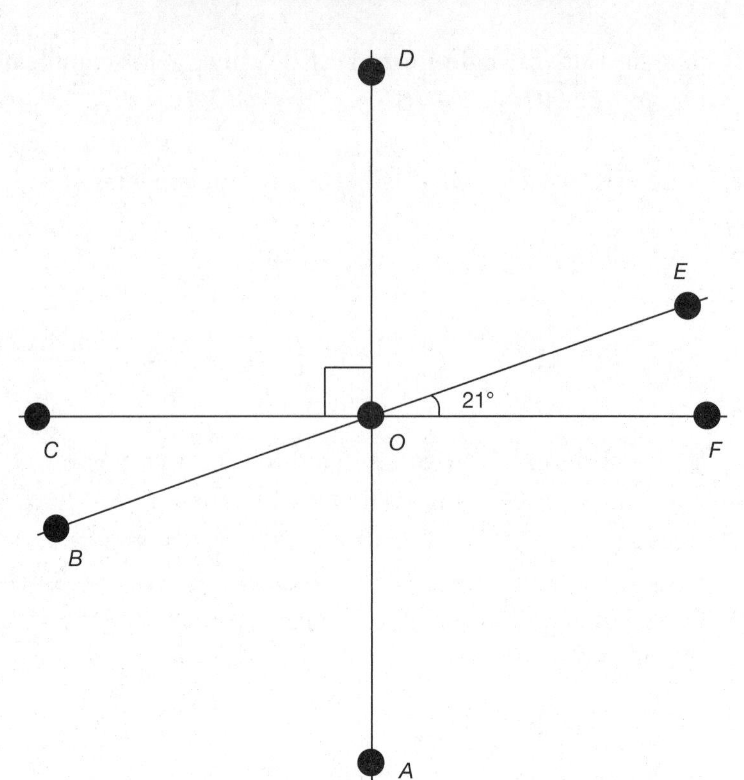

This diagram does not simply match the examples we saw along with the definitions of complementary, supplementary, and vertical angles. Now we will have to *find* the correct angle relationships before we can determine the angle measures, and we have to use those relationships to explain why the answers we get are correct. In a diagram like this, your child may be asked to find the measures of several different angles. The explanations below will give you models of how your child could justify his answers for angle relationship problems.

1. $m\angle DOE = 69°$ because $\angle DOE$ is complementary to $\angle EOF$, and $m\angle EOF = 21°$. $90° - 21° = 69°$.

2. $m\angle COB = 21°$ because $\angle COB$ is vertical to $\angle EOF$.

3. $m\angle BOF = 159°$. $\angle BOF$ is supplementary to $\angle EOF$, so $m\angle BOF = 180° - m\angle EOF = 180° - 21° = 159°$.

4. You can find $m\angle BOF$ lots of other ways. Here is one: $m\angle BOF = m\angle BOA + m\angle AOF$. $m\angle BOA = 69°$, since $\angle BOA$ is vertical to $\angle DOE$ (and we've already found $m\angle DOE$). $m\angle AOF = 90°$,

because $m\angle AOF$ is vertical to $\angle COD$, whose measure is 90° (given). Therefore, $m\angle BOF = m\angle BOA + m\angle AOF = 69° + 90° = 159°$.

That's a pretty complicated explanation for a pretty simple idea, but many students see angles that way, and that's okay—that's what makes it interesting, because you can solve it in your own way.

Parallel Lines and Transversals

In the diagram below, lines p and q are parallel—that is, they run in the same direction and will never meet, no matter how far you go along the lines (think of railroad tracks). Line t cuts across the two parallel lines and is called a **transversal**. In this diagram, there are 8 angles formed by the lines, which are labeled 1 through 8.

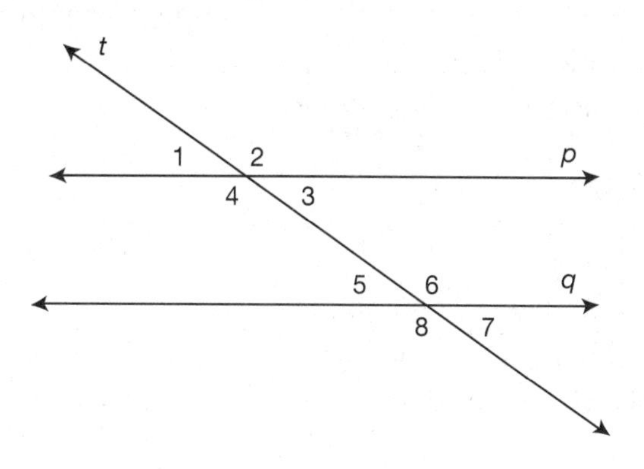

When a transversal cuts across two parallel lines, the two sets of four angles at each intersection are congruent (that is, equal) to each other. As you and your child will likely notice, there are lots of pairs of supplementary and vertical angles in this diagram.

TERMS TO REMEMBER

One main issue that gives most students trouble with these geometry topics is the vocabulary. There are a ton of new terms to learn here, and it will be important for you to help your child memorize the terms and to make sure he or she is using them properly. So far in this section on angle relationships, we have already learned so many words—*acute, right, obtuse, straight, complementary, supplementary, vertical.* We are about to learn some more that come up when parallel lines are cut by a transversal.

Corresponding angles are angles that are in the same position at the two different intersections. For example, $\angle 1$ and $\angle 5$ are corresponding because they are both in the upper-left corner of their respective intersections.

Corresponding angles are congruent to each other:

- $m\angle 1 = m\angle 5$
- $m\angle 2 = m\angle 6$
- $m\angle 3 = m\angle 7$
- $m\angle 4 = m\angle 8$

Alternate interior angles are angles that are *inside* the parallel lines, but on opposite sides of the transversal. For example, $\angle 3$ is alternate interior to $\angle 5$, and $\angle 4$ is alternate interior to $\angle 6$.

Alternate interior angles are congruent:

- $m\angle 3 = m\angle 5$
- $m\angle 4 = m\angle 6$

This is a very difficult relationship for some students to see. If your child is having trouble with it, try shading inside the angles with different colors to show which angles are alternate interior to each other.

Alternate exterior angles are angles that are *outside* the parallel lines and on opposite sides of the transversal. $\angle 1$ is alternate exterior to $\angle 7$, and $\angle 2$ is alternate exterior to $\angle 8$. Two alternate exterior angles are as far away from each other as possible, and point in opposite directions.

Alternate exterior angles are also congruent:

- $m\angle 1 = m\angle 7$
- $m\angle 2 = m\angle 8$

With alternate interior and alternate exterior angles, there is always another way to show their congruence. For example, we can show that $m\angle 2 = m\angle 8$ by reasoning that $\angle 2$ is corresponding to $\angle 6$, so $\angle 2$ and $\angle 6$ are congruent, and since $\angle 6$ is vertical to $\angle 8$, $\angle 8$ is congruent to both $\angle 6$ and $\angle 2$.

If two lines are *not* parallel, and they are cut by a transversal, their angle measures will not match up.

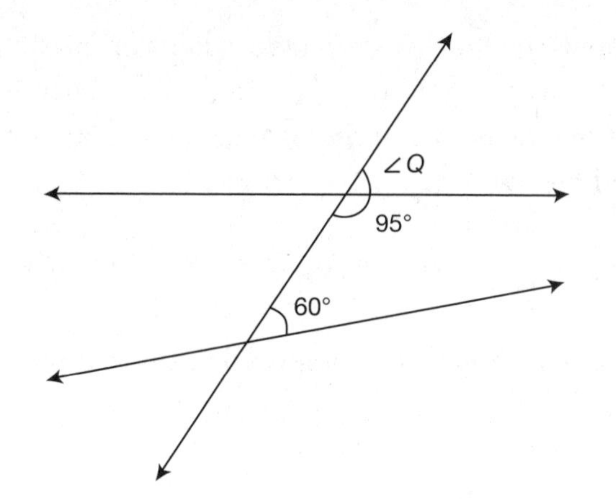

In this diagram, none of the lines are parallel. $\angle Q$ seems to correspond to the 60° interior angle. But $m\angle Q$ is not 60° because $\angle Q$ is supplementary to 95°. $m\angle Q = 85°$.

Angles of a Triangle

One of the most fun activities in middle school math is when students discover the relationship between the three angles of a triangle, which you can try at home. Draw a bunch of triangles—draw them large, and use a ruler to make sure the sides are straight. Trace an arc (part of a circle, so that it looks like someone took a bite off each corner (try to make the bites the same length, maybe an inch deep). Cut out the three corners, and place them together, point to point.

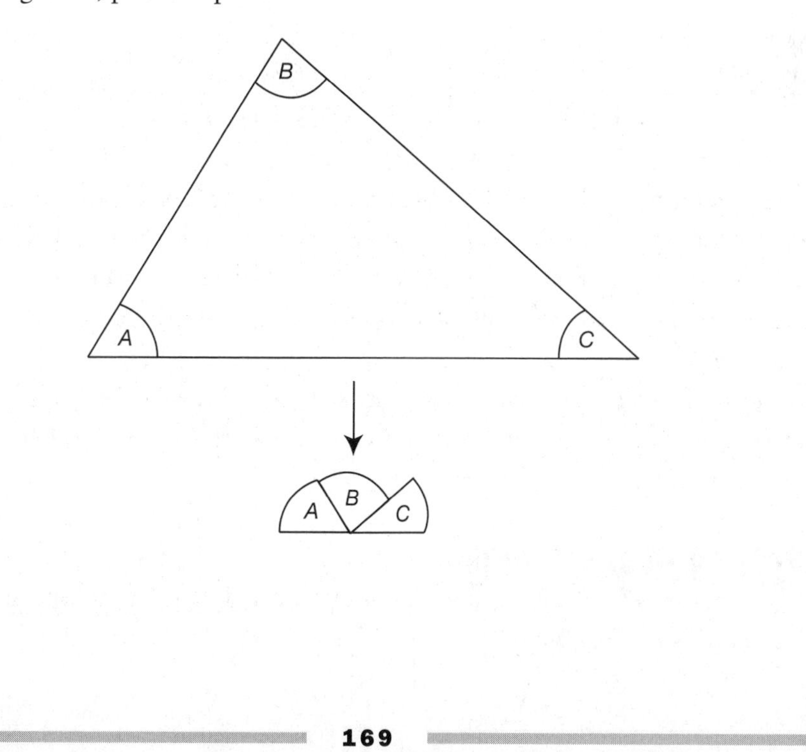

No matter how you draw your triangle, when you cut out the three corners, they will always fit together on a straight line. Therefore:

The sum of the measures of the three angles in any triangle is 180°.

This is called the **angle sum** of a triangle.

Using Transversals to Prove the Angle Sum of a Triangle

Now this is where math gets to be really cool. Look at the diagram below.

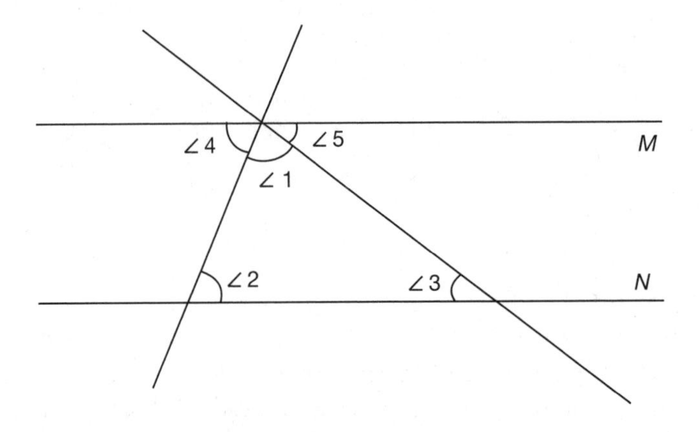

In this diagram, two parallel lines, *M* and *N*, are cut by two transversals. Both transversals cross the upper line at the same point, so that between lines *M* and *N*, a triangle is formed. ∠2 is alternate interior to ∠4, so they have the same measures. Similarly, ∠3 is alternate interior to ∠5.

Since $m\angle 4 = m\angle 2$ and $m\angle 3 = m\angle 5$:

$$m\angle 4 + m\angle 1 + m\angle 5 = m\angle 2 + m\angle 1 + m\angle 3$$

That's a lot of symbols, so let's break it down. The left side of the equation shows the sum of the three angles along the underside of line *M*. The right side of the equation shows the sum of the three angles inside the triangle. Since the angles along a line add up to 180°, so do the angles inside the triangle.

If this doesn't immediately make sense to you, it would be a great time to have some fun doing math with your child—work together to make sense of this argument.

Solving Angle Problems

Now that we know a lot about angles and how they relate to one another, let's solve some problems.

1. Find the measure of the unknown angle in the triangle.

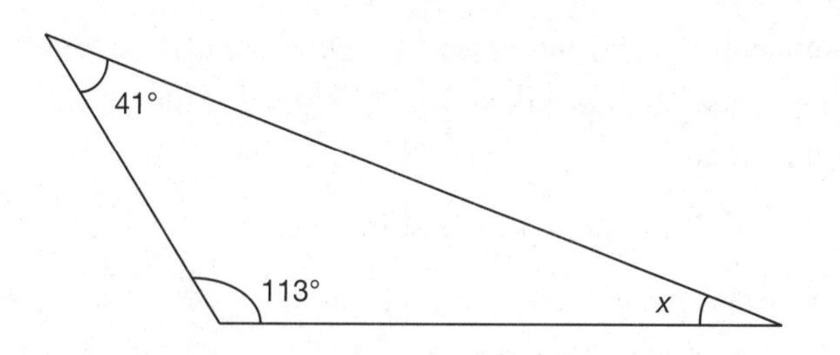

We know that the measures of the angles inside a triangle always add up to 180°. The two known angles have measures of 41° and 113°. So we can subtract the two known angle measures from 180°: $180° - 41° - 113° = 26°$. $m\angle x = 26°$

2. Find the measure of angle y.

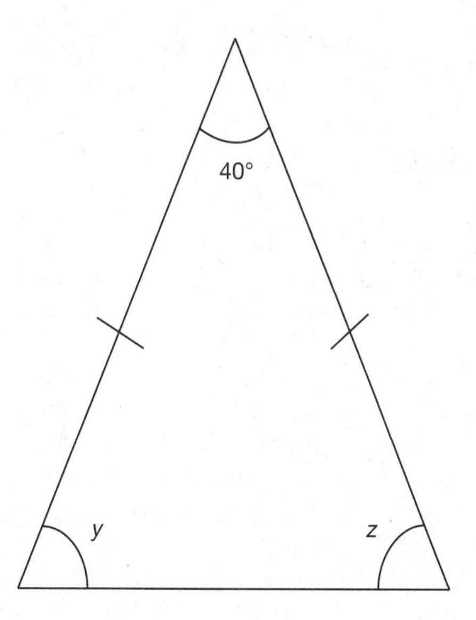

In this triangle, we only know one angle measure. But there is another clue. The little marks through the two diagonal sides of the triangle tell us that those sides are congruent—they are the same length. That means this is an isosceles triangle, and that two of its angles have the same measure. Angles y and z are the congruent angles because they are opposite the congruent sides. So $m\angle z = m\angle y$. If these two angle measures are the same, then all we need to do is subtract the known angle measure from 180° to find out how many degrees are shared by these two angles, then divide by 2. $180° - 40° = 140°$. $140° \div 2 = 70°$. $m\angle y = 70°$.

ERROR ALERT!

Did you notice in the last paragraph that one equation ended with 140°, and the next equation started with 140°? Many students would write the two steps like this:

$$180° - 40° = 140° \div 2 = 70°$$

This is what many math teachers call a *run-on equation*. Even though this method gives you the correct answer of 70°, it is not a correct way to write an equation. In an equation, the expressions on either side of the equals sign must have the same value. If that were true here, then 180° — 40° would be the same as 70°, which it is not. If your child is performing multiple steps, make sure she is not writing them one after another like this. A better way to show this work would be to start a new line for each step:

$$180° - 40° = 140°$$

$$140° \div 2 = 70°$$

3. Find $m\angle MNP$.

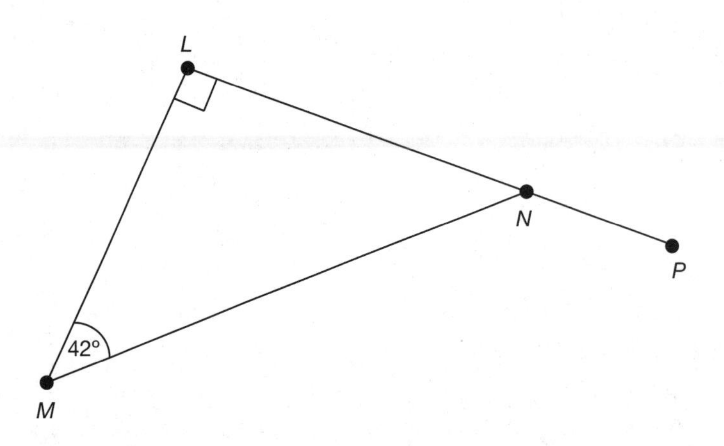

Angles $\angle MNL$ and $\angle MNP$ are supplementary. If we can find the measure of $\angle MNL$, we're almost there. Since $\angle MNL$ is inside the triangle, we can use the other two known angles to find it. Some students might be thrown off by the fact that $\angle MLN$ has no number given, but we know that it is a right angle because of the box in the corner. Therefore

$m\angle MNL = 180° - 90° - 42°$
$m\angle MNL = 48°$

Since $\angle MNL$ and $\angle MNP$ are supplementary:

$m\angle MNP = 180° - m\angle MNL$
$m\angle MNP = 180° - 48°$
$m\angle MNP = 132°$

4. In the diagram below, $m\angle t = 37°$. Find $m\angle s$.

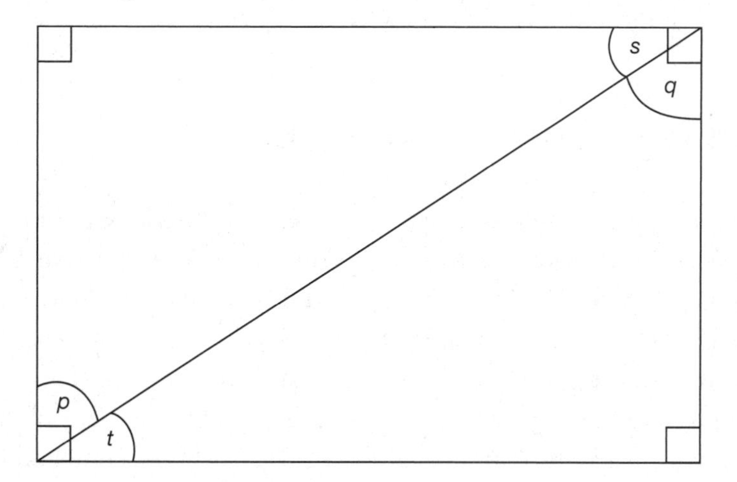

This one may seem easy, but it's all about how you explain your answer. There are many ways to do it. Here are just two.

First, since the outer shape is a rectangle (it has four right angles), the top and bottom sides are parallel. Therefore, the line connecting angles s and t is a transversal, and $\angle s$ is alternate interior to $\angle t$. $m\angle s = m\angle t$, so $m\angle s = 37°$.

Another way is to see that angles t and p are complementary. Therefore, $m\angle p = 90° - 37° = 53°$. Angles p, s, and the right angle in the upper-left corner form a triangle, so their measures add up to 180°.

$m\angle s = 180° - 90° - m\angle p$
$m\angle s = 180° - 90° - 53°$
$m\angle s = 37°$

Students enjoy solving problems with angles because they require a certain amount of visual thinking and problem solving, and the arithmetic is simple. You may find that angles are the first math concept your child has enjoyed in awhile. If so, take advantage of the situation. Use it as an opportunity to build your child's confidence and maybe even practice some mental math (on all that subtraction and the occasional division). Let your child know that he is doing well and that you believe he can be successful in math if he continues to work at it.

The Pythagorean Theorem

One of the most famous, and fun, concepts in middle school math is the Pythagorean theorem. It is often referenced in popular culture, from *The Wizard of Oz* to *The Simpsons*, and YouTube is full of examples and proofs. The Pythagorean theorem is an idea that has been explained and proven in hundreds of ways, including one proof by U.S. President James Garfield. You may not know it, but you use the Pythagorean theorem all the time—when you take a shortcut through your neighborhood, hang holiday lights, climb a ladder, or buy a new TV.

The importance of the Pythagorean theorem in middle school math cannot be overstated. While it is a relatively simple idea, it unites algebra and geometry in a way that is critical to building a student's foundation for high school. It gives students lots of practice with some fundamentals of algebra, including substitution, exponents, and square roots. It is also an opportunity for students to deepen their understanding of numbers, including irrational numbers, and to develop a sense of the distinction between exact answers and approximations. And it is one of the basic ideas on which physics, calculus, and other higher-level math and science courses will depend.

Eighth-grade students love the Pythagorean theorem. I've taught 8th grade for years, and year after year, no matter what group, I have witnessed my students almost unanimously lighting up when it comes to this topic. Often the students who have been disengaged all year step up and become leaders in the class when we hit this topic. The algebra looks fancy, and kids get to use sophisticated-sounding words like *hypotenuse*. The problems are always visual and can be related to almost any topic. And there's a special feeling when the numbers add up to a perfect square. All of these combine into a mysterious attraction for students—they want to solve problems like never before.

Explaining the Theorem

As you may remember, the Pythagorean theorem concerns the relationship between the three sides of a right triangle. A **right triangle** is any triangle that has one 90° angle. The two sides of the triangle that meet to form the right angle are called the **legs** of a right triangle. The third side, which is always the longest, is called the **hypotenuse**.

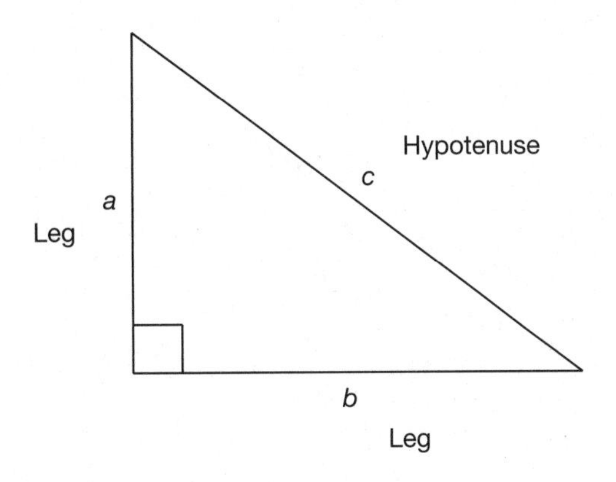

We use the variables a and b to represent the lengths of the two legs. It does not matter which is which. However, we always use c to represent the length of the hypotenuse.

THINK ABOUT IT TOGETHER

An interesting question to ask your child is whether a triangle can have *two* right angles. What is the angle sum of a triangle? What happens if you try to draw a triangle with two right angles?

The Pythagorean theorem simply states that in any right triangle, the following relationship is true:

$$a^2 + b^2 = c^2$$

To put it in words, if you create squares with side lengths a, b, and c, the squares of the two legs combined are equal (in area) to the square of the hypotenuse.

Here is a visual that may help conceptualize it:

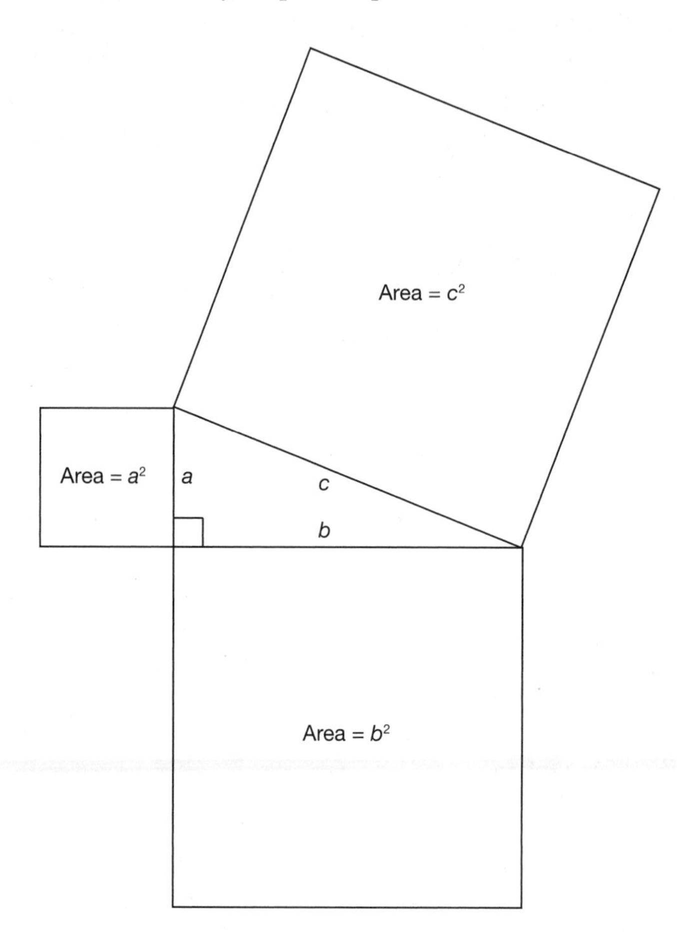

The tilted square, built off the hypotenuse, is the largest of the three squares, and its area is equal to the sum of the areas of the other two squares.

There are many ways of proving this theorem, and everybody has different preferences as to which makes the most sense. Your child's teacher will likely show one or two of these proofs in class, and it would be worthwhile to ask your child to explain the proof to you. Many students are able to follow the proof in class, but have a hard time paraphrasing it, usually because it involves a bit of high school algebra. Don't worry if your child cannot explain the proof at this point. Just discussing it is a great way to help make sense of the concept. This is also a great time to check YouTube, which has many visual proofs of the Pythagorean theorem.

The Classic Example

The most famous example of the Pythagorean theorem is the 3-4-5 right triangle:

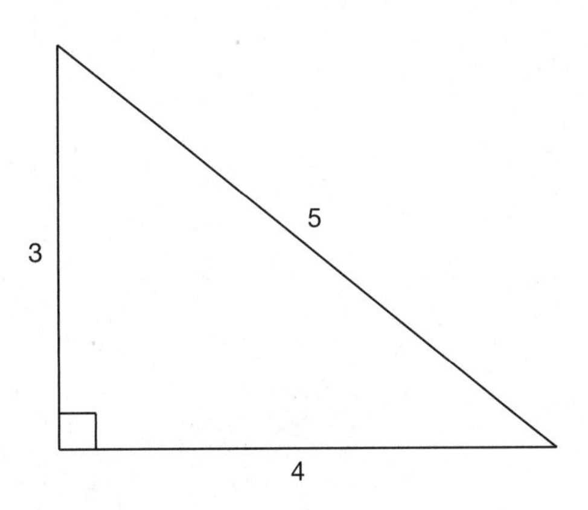

In this example, the lengths of the legs are 3 and 4 units, and the length of the hypotenuse is 5 units. So $a = 3$, $b = 4$, and $c = 5$. Let's see how the equation works out:

$$a^2 + b^2 = c^2$$

$$3^2 + 4^2 = 5^2$$

Now we evaluate all the squares:

$$9 + 16 = 25$$

And finally, add up the squares of the legs:

$$25 = 25$$

Since the numbers add up correctly, the relationship holds true. In words, what we have shown is that the areas of the squares off the legs add up to the area of the square off the hypotenuse. The legs are 3 and 4, so their squares have areas of 9 and 16. The hypotenuse is 5, so its square has an area of 25: 9 + 16 = 25.

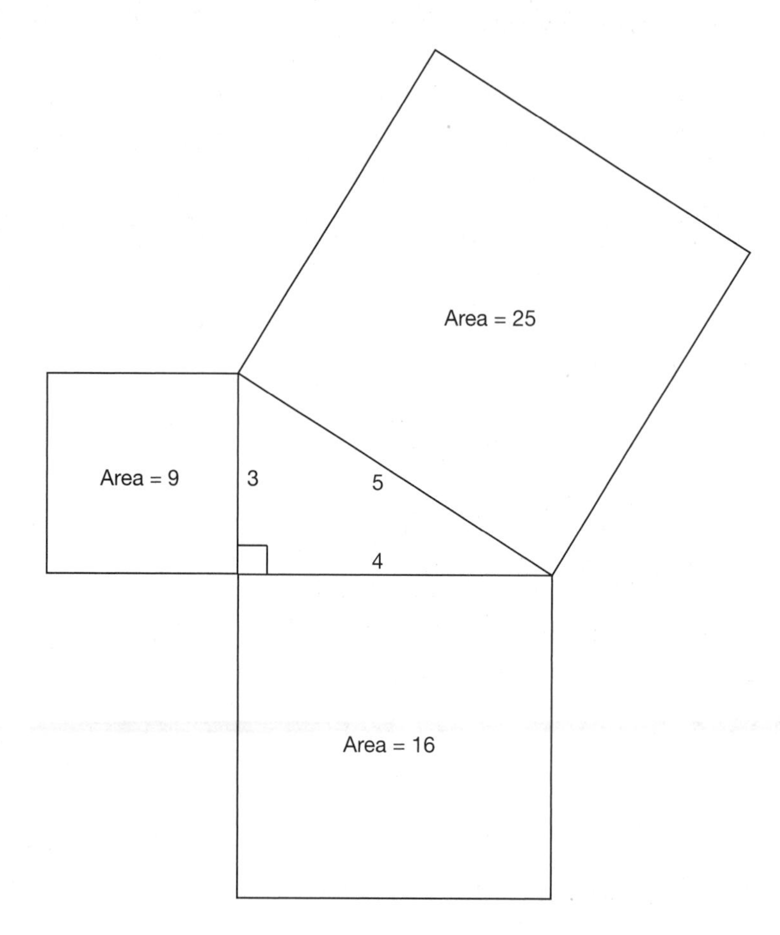

MISCONCEPTION ALERT!

In the pictures for the 3-4-5 example, I have drawn the triangle with the same orientation each time. Many teachers and textbooks always show right triangles with the right angle in the same location, so that the hypotenuse is always diagonal, and students become accustomed to seeing the right angle in a particular place, like the lower-left corner. This gives students a misconception that the hypotenuse *must* be diagonal, and right triangles must *always* look that way. But a right triangle could be rotated to any orientation. One common error is not recognizing that a right triangle, such as the one below, is a right triangle.

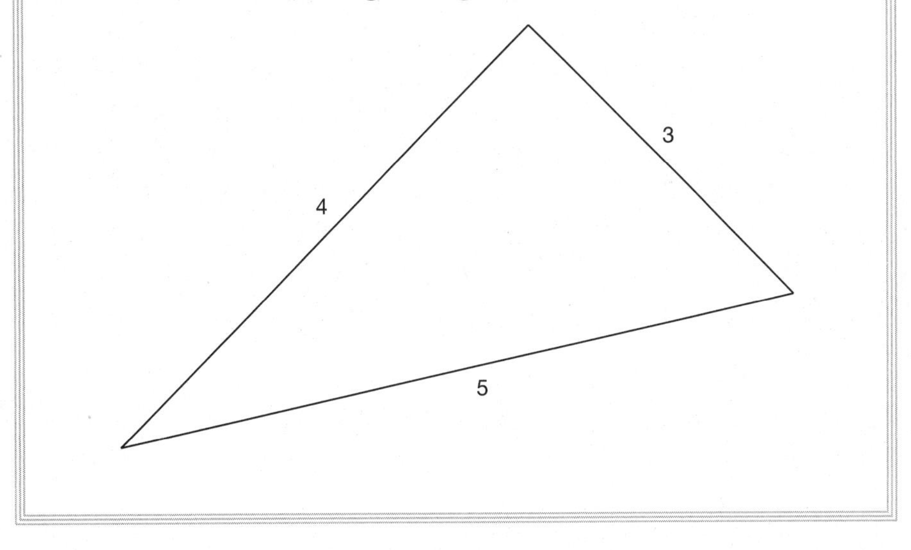

In Chapter 5, we worked with scale drawings. If you were to scale a 3-4-5 right triangle by any factor, you would still have a right triangle. For example:

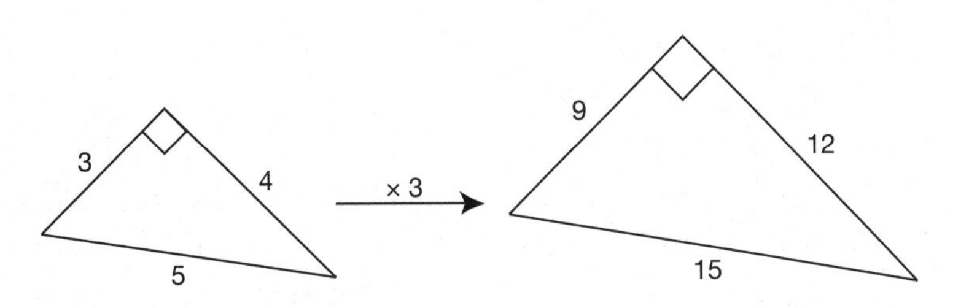

The larger triangle, which has side lengths 9, 12, and 15, is still a right triangle, and still satisfies the Pythagorean theorem:

$$a^2 + b^2 = c^2$$

$$9^2 + 12^2 = 15^2$$

$$81 + 144 = 225$$

$$225 = 225$$

ORIGIN OF THE THEOREM

Pythagoras was a famous Greek mathematician, philosopher, and explorer. In some tellings of the story, Pythagoras discovered the theorem when he was traveling in Egypt. He saw people using a loop of rope with 12 knots to make right angles. The 12 knots were equally spaced, so three people could pull the rope taut to make a 3-4-5 triangle, which always has a right angle.

The Converse of the Pythagorean Theorem

Converse is a fancy math term for the opposite of a statement. The Pythagorean theorem says that any right triangle has the relationship $a^2 + b^2 = c^2$. The converse of the Pythagorean theorem says that any three lengths that satisfy that relationship will form a right triangle.

Examples

1. Will three beams of lengths 7 feet, 9 feet, and 14 feet form a right triangle?

In order to find out, we can simply plug the numbers into the equation. It does not matter which length we use for a and which we use for b, but we must *always* use the largest of the three numbers for c. As we substitute our lengths into the equation, we write a "?" on top of the equals sign until we know for sure whether the equation is true.

$$a^2 + b^2 = c^2$$
$$7^2 + 9^2 \stackrel{?}{=} 14^2$$
$$49 + 81 \stackrel{?}{=} 196$$
$$130 \neq 196$$

Since the squares of the legs do not add up to the square of the hypotenuse, three beams with lengths 7, 9, and 14 will *not* form a right triangle.

2. Do the lengths 6, 8, and 10 form a right triangle?

$$a^2 + b^2 = c^2$$
$$6^2 + 8^2 \stackrel{?}{=} 10^2$$
$$36 + 64 \stackrel{?}{=} 100$$
$$100 = 100$$

Since $36 + 64 = 100$, these three numbers make a right triangle. They are called a **Pythagorean triple**. You may have already noticed that 6-8-10 is proportional to 3-4-5 by a scale factor of 2. Therefore, it is a similar triangle with a right angle—it has to satisfy the Pythagorean theorem.

NOTE

As your child is working on the Pythagorean theorem, it is a good time to memorize the squares of all the whole numbers up to at least 17. This will help immensely as your child works through numerous examples where he/she has to calculate squares and square roots.

Solving Problems with the Pythagorean Theorem

The great thing about the Pythagorean theorem is that it allows us to calculate lengths and distances that we may not be able to measure directly.

Computing the Hypotenuse

1. In the picture below, a cat is stuck in a tree 12 meters above the ground. A brave firefighter will attempt to rescue the cat by placing her ladder 5 meters away from the tree. How long does the ladder have to be?

THE MEANING OF "REAL LIFE"

This is an example of how math teachers can abuse the term *real life*. In real life, the firefighter wouldn't be constrained to place the ladder 5 meters away from the tree, and she certainly wouldn't pull out a tape measure to see exactly how high the cat is. She would have her ladder, which is a certain size, and she would place it where it needs to be. In the process, she might do some quick mental estimation, which might be related to the Pythagorean theorem, but she is unlikely to be thinking about square roots in this situation—and there is no need for it.

This isn't to say that the Pythagorean theorem isn't important or relevant to real-life situations, but you should help your child see that the contextual problems presented by teachers and textbooks are often contrived to fit the math, rather than to truly be realistic.

We can use the Pythagorean theorem to find out how long the ladder needs to be.

We know the lengths of the two legs of a right triangle, but we do not know the length of the hypotenuse. So we will leave c as a variable until we figure it out.

$$a^2 + b^2 = c^2$$

First we substitute the values we know: $5^2 + 12^2 = c^2$

Now evaluate the squares: $25 + 144 = c^2$

And add: $169 = c^2$

Now we take the square root of both sides: $\sqrt{169} = \sqrt{c^2}$

$$13 = c$$

So the ladder needs to be 13 meters long.

ABOUT NOTATION

A very common error that students make is in how they express their answer to a problem like this. After computing the square root, many students are not clear about the notation. They may write any of the following INCORRECT answers:

$$c^2 = 13$$

$$c = 13^2$$

$$c = \sqrt{13}$$

$$\sqrt{c} = 13$$

The answer to this problem should be expressed as $c = 13$ because we have evaluated the square root of both c^2 and 169. There should not be any exponents or square root signs left over when you get to your final answer.

In the last example, the numbers worked out nicely, where the hypotenuse ended up being a whole number. This will often not be the case, such as in this example:

2. Find the length of the hypotenuse of a right triangle whose legs are 5 and 6 units long.

	$a^2 + b^2 = c^2$
First we substitute the values we know:	$5^2 + 6^2 = c^2$
Now evaluate the squares:	$25 + 36 = c^2$
And add:	$61 = c^2$
Now take the square root:	$c = \sqrt{61}$
	$c \approx 7.8$

The last step is one that gives many students trouble. Your child may learn methods for estimating this mentally, which is a skill worth learning. Since 61 is close to 64, which is 8^2, we can estimate that c is just a little bit less than 8. A calculator will give you

$$\sqrt{61} \approx 7.81024968$$

Note that in our answer, we have used a squiggly equals sign (≈). This symbol indicates that our answer is approximate. If you find $\sqrt{61}$ on a calculator, you will see that there are many digits to that decimal—in fact, it goes on forever with no repeated pattern (more on this later in the chapter). Nobody needs all those digits, so we round our answer, in this case to 7.8. To show that we are rounding, we need to use ≈.

Many students, when taking the square root, will mistakenly find *half* of the number, for example, answering 30.5 in the previous example. If your child is making this mistake, help him/her see that the answer does not make sense. If you were to draw a triangle with side lengths 5, 6, and 30.5 (or 30, just to use a nice number), you would see why that answer is unreasonable.

So in summary, when you know the lengths of the two legs and you need to find the length of the hypotenuse, use the Pythagorean theorem: Square both legs and add the squares together to find the square of the hypotenuse, then find the square root of that sum.

Computing a Leg

Mohammed is buying a new TV for his living room. He has limited space, so he wants to make sure the TV will fit. When TV measurements are given, the number given is the length of the *diagonal* of the TV—from one corner to the opposite corner. A 42-inch TV has a height of 20.5 inches. How wide is the TV?

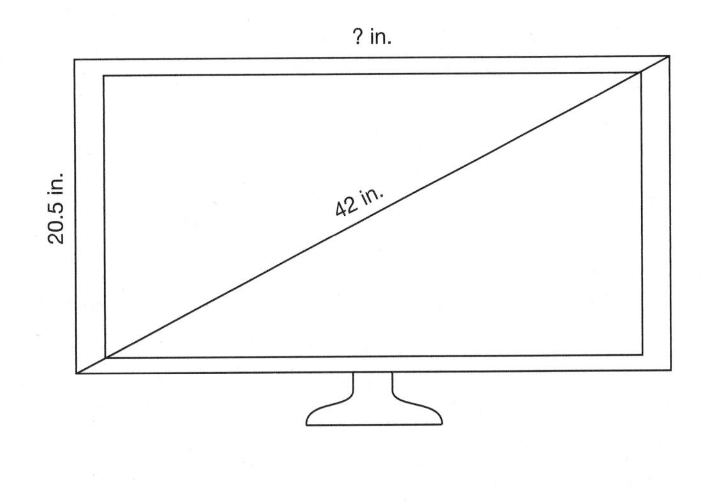

We can cut the TV in half along the diagonal to form two right triangles. We know the lengths of the hypotenuse and one leg. Our unknown value is one of the legs. Let's call it b.

	$a^2 + b^2 = c^2$
Substitute the known values:	$20.5^2 + b^2 = 42^2$
Evaluate the squares:	$420.25 + b^2 = 1{,}764$
Subtract 420.25 from both sides to isolate b^2:	$b^2 = 1{,}764 - 420.25 = 1{,}343.75$
Take the square root of both sides:	$b = \sqrt{1{,}343.75}$
Evaluate the square root:	$b \approx 36.7$ inches

So the TV would be approximately 3 feet (36 inches) in width. If the algebra used here does not make sense to you right now, do not worry. We will come back to it in Chapter 8.

Finding Distances on a Grid

Another use of the Pythagorean theorem is finding distances on a coordinate plane or any grid. If two points are on a grid, you can use the grid to create a right triangle, and find the length of the hypotenuse to figure out how far apart the two points are.

Examples

1. How far apart are the points (1,3) and (5,6)?

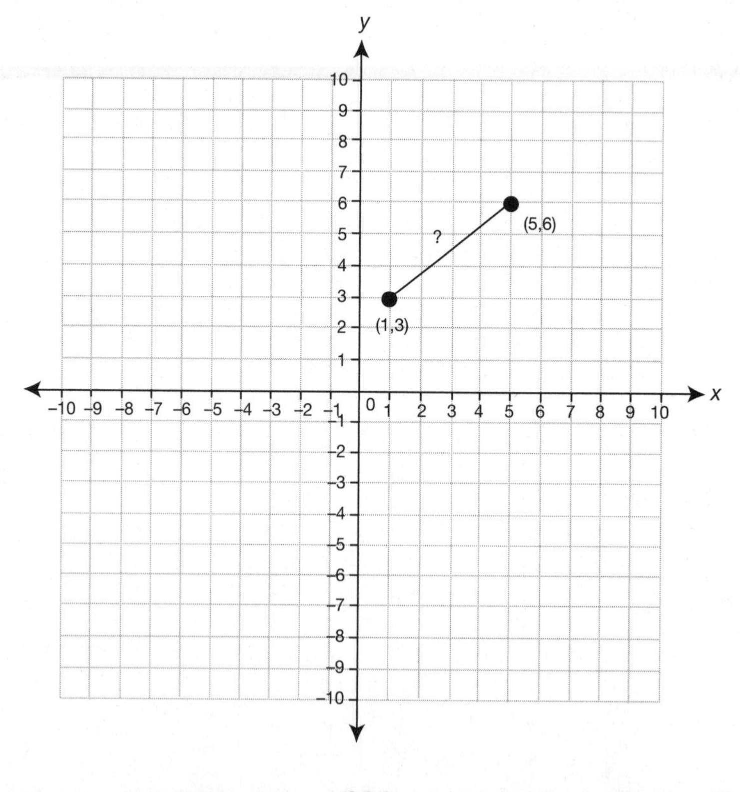

The straight line between the points is diagonal. Many students will attempt to count the boxes that this line passes through. This is not a correct way of solving the problem, because the line cuts a different length through each box. Instead, we can create a right triangle:

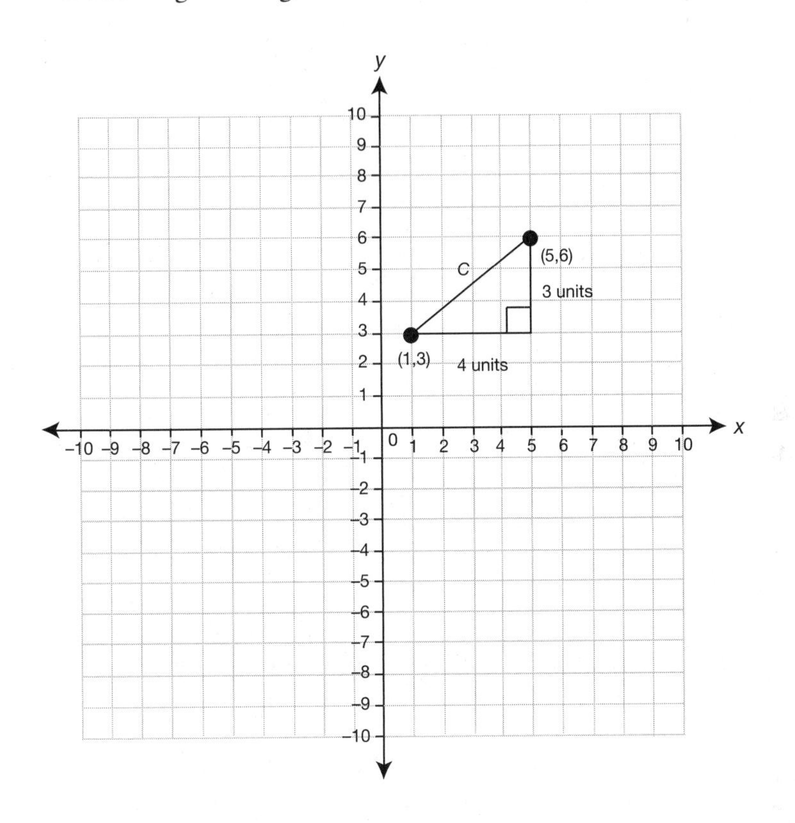

Remember the 3-4-5 right triangle? Here it is again. Since the horizontal leg is 4 units long and the vertical leg is 3 units long, the hypotenuse is 5 units long. Therefore, the points are 5 units apart. When you recognize a 3-4-5 right triangle, you don't even need to calculate anything!

2. How far apart are points P and Q in the graph below?

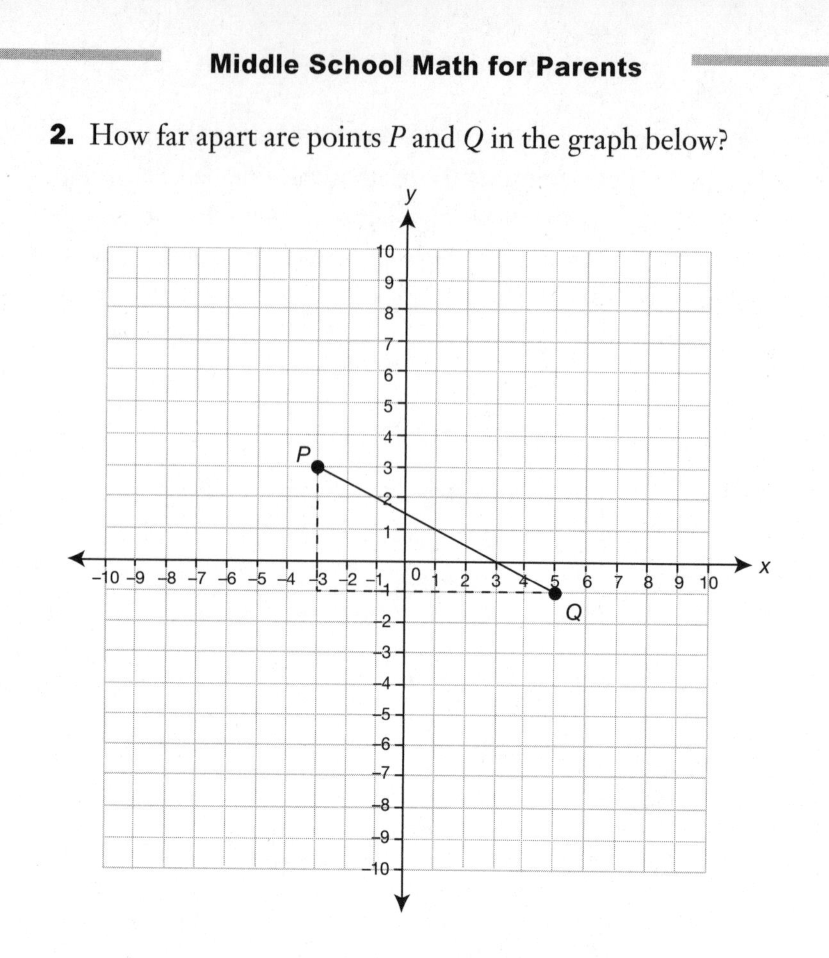

In this diagram, I have used dotted lines to show the legs of the right triangle. The vertical leg is 4 units long, and the horizontal leg is 8 units long. So the distance (c) is:

$$a^2 + b^2 = c^2$$
$$4^2 + 8^2 = c^2$$
$$16 + 64 = c^2$$
$$80 = c^2$$
$$c = \sqrt{80} \approx 8.9$$

So points P and Q are roughly 8.9 units apart.

Transformations

Another topic that students enjoy a great deal is **transformation**. In geometry, a transformation is a way of changing a figure. There are four types of transformations that your child will learn in middle school: **translations**, which slide a figure in a specific direction; **reflections**, which flip a figure across a mirror, or line of reflection; **rotations**, which change the orientation and position of a figure by turning it around a certain point; and **dilations**, which stretch or shrink a figure by a certain scale factor.

If your child is artistic, enjoys crafts, plays video games, or is interested in video animation, then he will likely find this topic exciting. Transformations apply to all of these activities, as well as many others, and appeal to most visual learners.

Translations

In geometry, to translate something means to move it without changing its size or orientation. In the real world, you don't often see a pure translation because things change as they move. You cannot move your body if your body is perfectly still, and a car cannot drive without its wheels rotating. Think of when your child was small and would go down a slide. His body could be mostly still as he moved down the slide. This is a good example of a translation.

In geometry, we use the most precise language possible to describe translations. When we translate a figure on a coordinate plane, we always describe the translation by how far and in what direction the figure moved.

Example

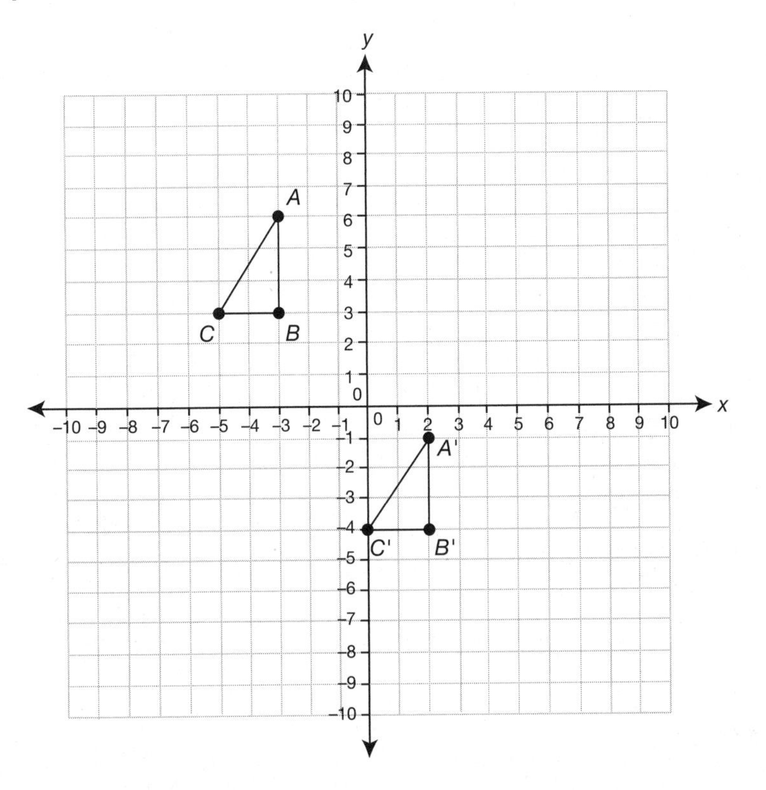

In this diagram, triangle *ABC* has been translated 5 units to the right and 7 units down. If you place your pencil on any of the original vertices, *A*, *B*, or *C*, and move 5 units right and 7 units down, your pencil will end up on the image point *A*', *B*', or *C*', respectively. Both triangles are exactly the same size, and face the same way.

When asked to describe a translation that has occurred, your child should choose one vertex of the figure (such as point *A*) and count how far it moved in each direction.

ABOUT NOTATION

When a transformation occurs, we need a way to identify which figure came first. For example, if triangle *ABC* is translated, then *ABC* is referred to as the original figure. The new triangle after the translation would be *A'B'C'*, and we say it is the image of triangle *ABC* under a translation. The ' symbol is called prime.

Reflections

A reflection is just what it sounds like: a mirror image. When we reflect figures geometrically, we use a line as a mirror. At this level, we will focus on reflections in the x-axis and reflections in the y-axis.

Examples

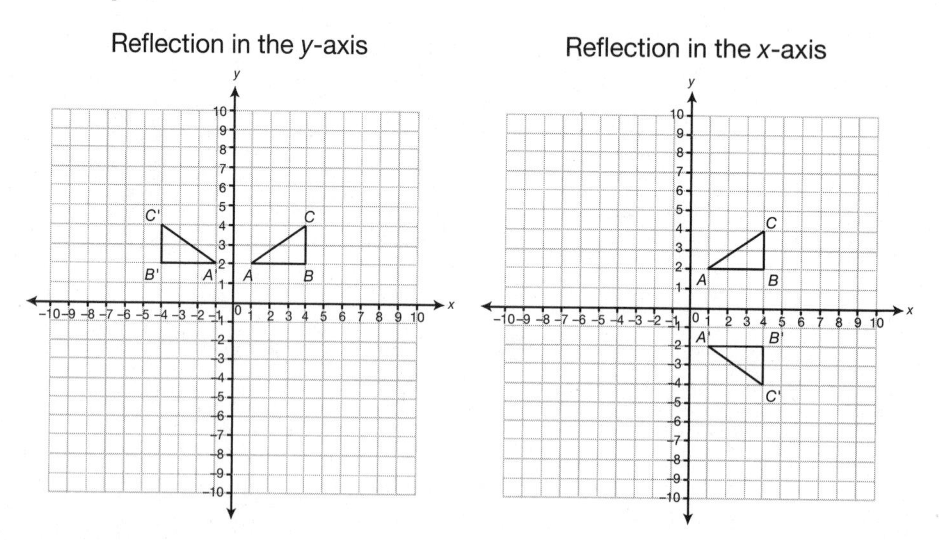

When a figure is reflected in the y-axis, it flips horizontally (left and right reverse). When a figure is reflected in the x-axis, it flips vertically (upside-down). Notice that when the triangle is reflected in the x-axis, it moves down. This is not a translation—that movement is a feature of the reflection that has occurred. If you look at a mirror and take a step toward the mirror, your mirror image also moves closer to the mirror. If you take a step away from the mirror, your image also moves away from the mirror. A figure's image under a reflection will end up on the other side of the line of reflection, the same distance from the line of reflection as the original figure.

MISCONCEPTION ALERT!

Because a reflection in the *x*-axis seems to move a figure up or down, many students mistakenly say the figure is reflected in the *y*-axis, since the *y*-axis points up and down. Similarly, when a figure is reflected in the *y*-axis and moves either left or right, many students mistakenly say the figure is reflected in the *x*-axis. You can help your child with this misconception by reminding her that a reflection occurs in a mirror, and pressing her to name what is acting as the mirror in each problem. A reflection in the *x*-axis is a reflection that uses the *x*-axis as its mirror.

If you start with your pencil on any point of the triangle and count boxes to the line of reflection, you can count the same number of boxes *past* the line of reflection to find the image point.

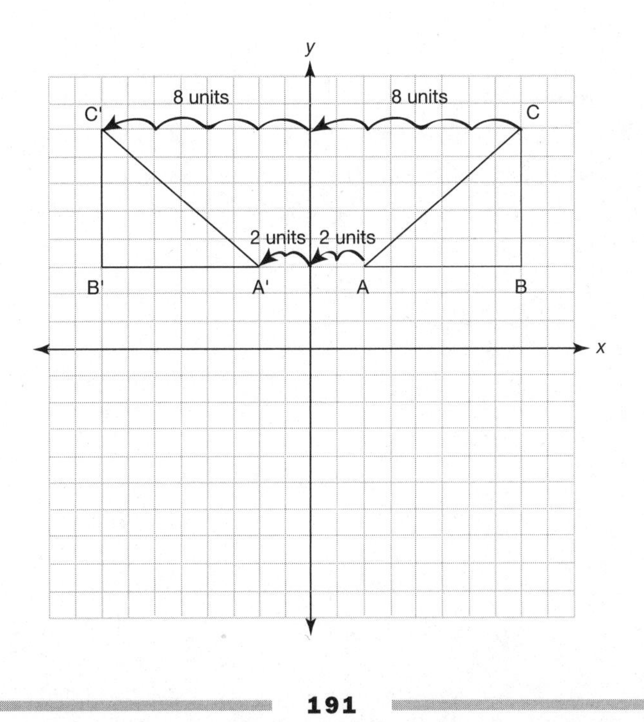

You or your child may also notice that the coordinates of the image points are just like the coordinates of the original points, with one difference: One of the coordinates is reversed (opposite sign). Reflections in the x-axis change the y-coordinate, and vice versa.

Sometimes students have trouble when asked to reflect figures across a line that passes through the figure. For example, suppose we are trying to reflect the following triangle in the x-axis:

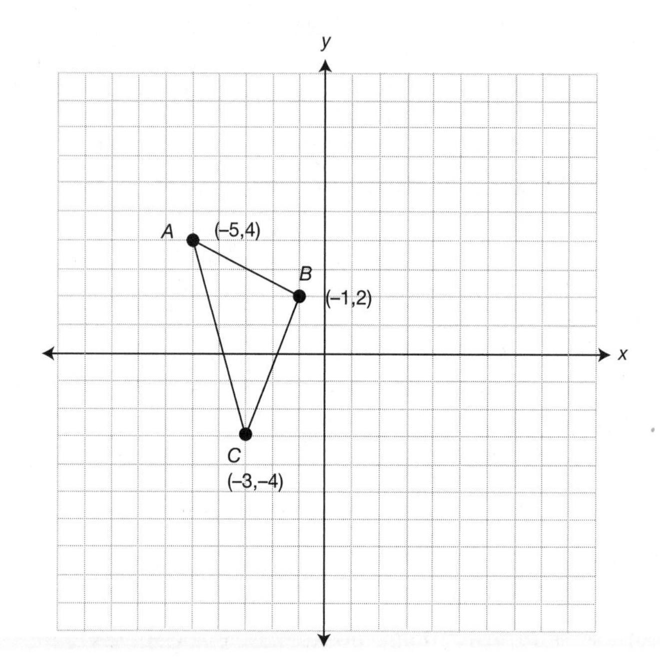

Since the x-axis passes through this triangle, many students have trouble with this reflection, and often they think it cannot be done. However, we can use the same idea we used in the previous example. If we count each point's distance to the mirror and go just as far in the other direction, we can reflect the triangle point by point:

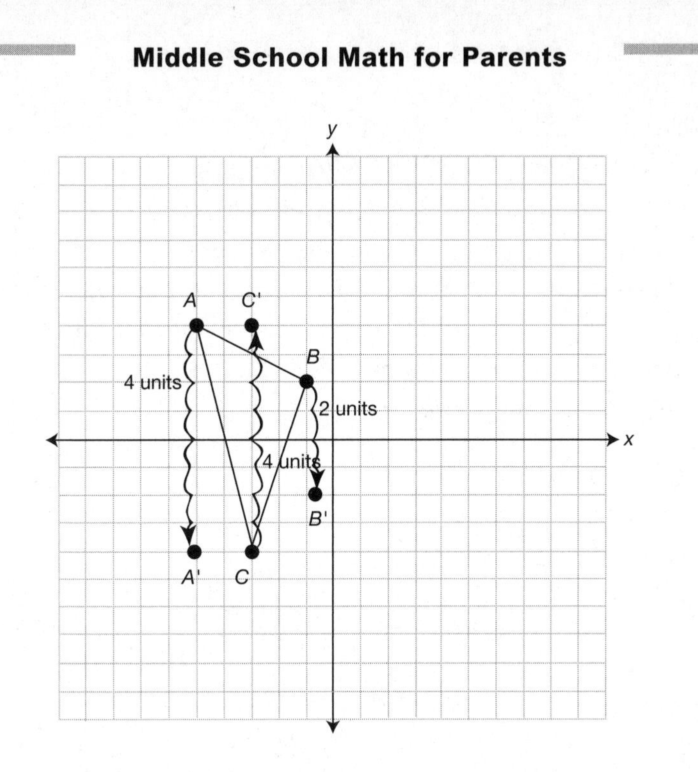

And now we can draw the new triangle, which will overlap with the original triangle:

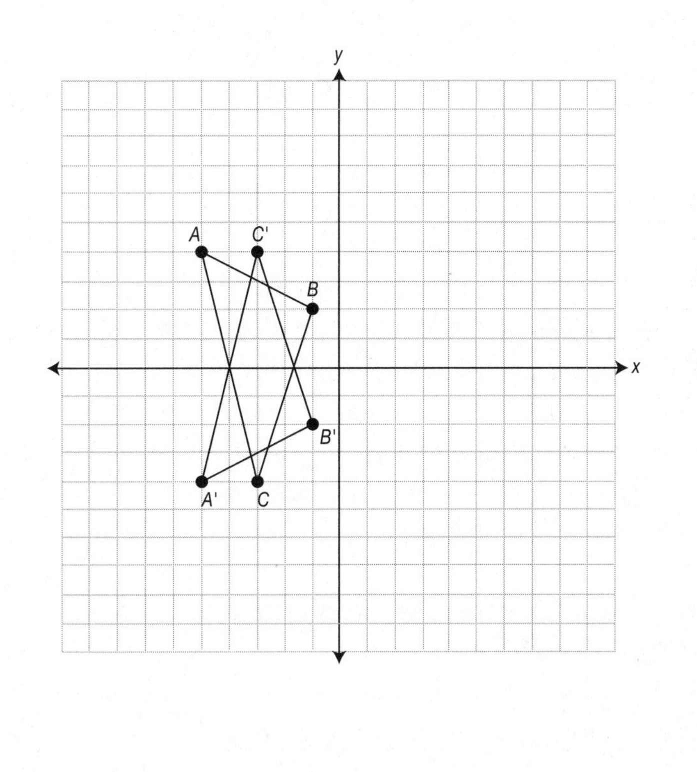

Rotations

Rotations are the most difficult transformation for most students. When the wheels of a car rotate, they turn around an axle. When a geometric figure rotates, it must rotate around a center point. In most cases, your child will be asked to perform rotations with the origin (0,0) as the center of rotation. Rotations also need to have a certain angle of rotation. For our purposes, we will focus on rotations by 90°. You can get a rotation of 180°, 270°, or 360° by repeating a 90° rotation.

Example

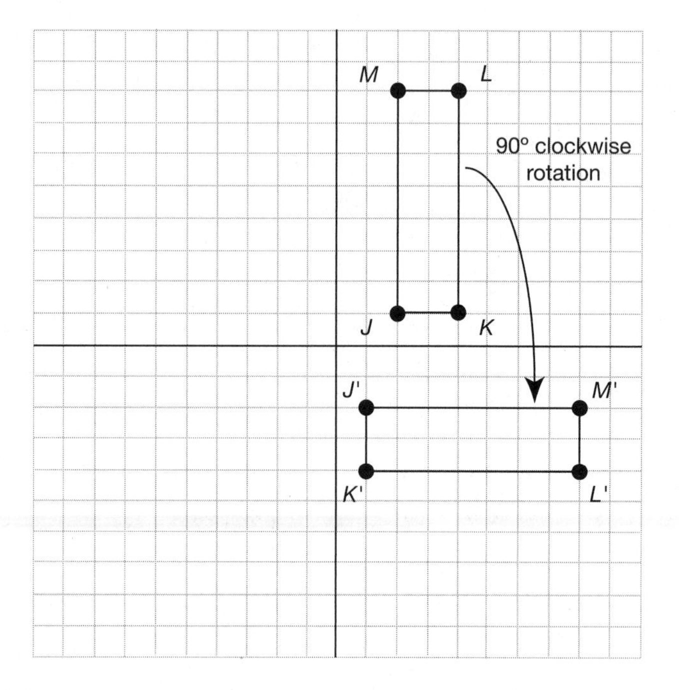

In this example, rectangle *JKLM* has been rotated 90° clockwise about the origin. The new rectangle, *J'K'L'M'* is in a different quadrant of the graph, and is oriented a different way.

One way to see rotation is to press the tip of your pencil into the origin and spin the paper 90°. Then you can look at where the figure ends up and draw it there on the original graph.

Congruence Transformations and Similarity Transformations

Translations, reflections, and rotations all create new images that are identical to the original figure in size and shape. In geometry, two figures that have the same shape and size are called **congruent**, so translations, reflections, and rotations are all considered **congruence transformations**.

In Chapter 5, we saw that scaling changes the size of a figure but keeps it similar to the original. The name for this kind of transformation is a **dilation**. A dilation creates a similar figure, not a congruent one, because dilations change the size of a figure. On a coordinate grid, you can perform a dilation by multiplying the coordinates of each point by the scale factor. If the scale factor is greater than 1, this moves the points farther away from the origin. If the scale factor is between 0 and 1, it moves the points closer to the origin.

Examples

1. Here is an example of a dilation by a factor of 2:

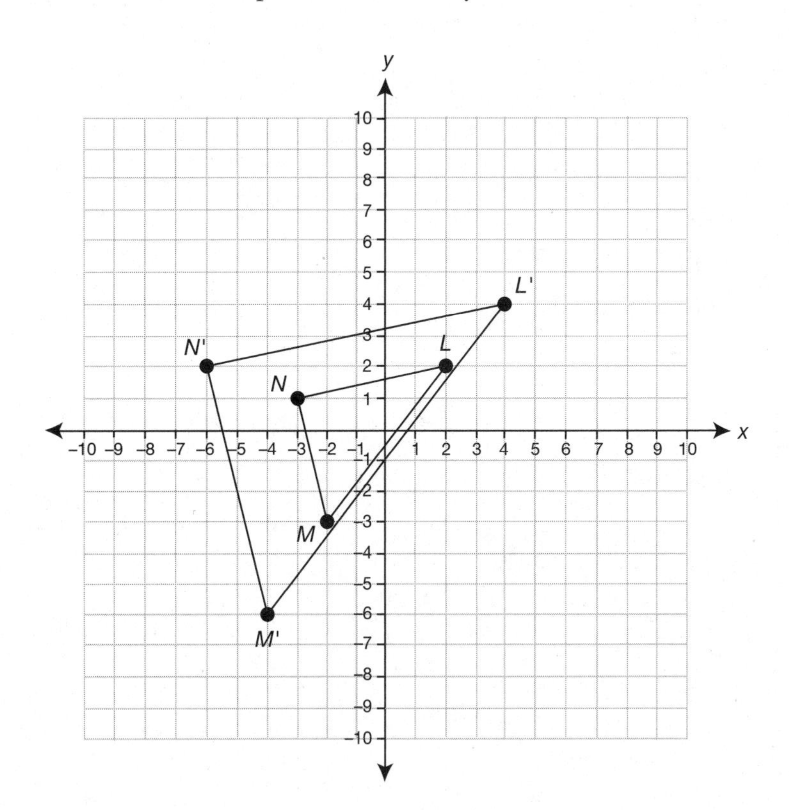

The image triangle's dimensions are double the original triangle's dimensions. Notice that if you draw a line segment connecting the origin to point *L'*, *M'*, or *N'*, that line will pass through the original point *L*, *M*, or *N*—in fact, since the scale factor was 2, the original point would be the midpoint of the line segment.

In this example, all of the coordinates were doubled, as shown in the table below.

Original Point	Image Point
L(2,2)	*L*'(4,4)
M(−2,−3)	*M*'(−4,−6)
N(−3,1)	*N*'(−6,2)

2. When the original figure is wholly contained within one quadrant, it can be easier to see how the shape moves away from or toward the origin, as in this example of a dilation by a factor of 3:

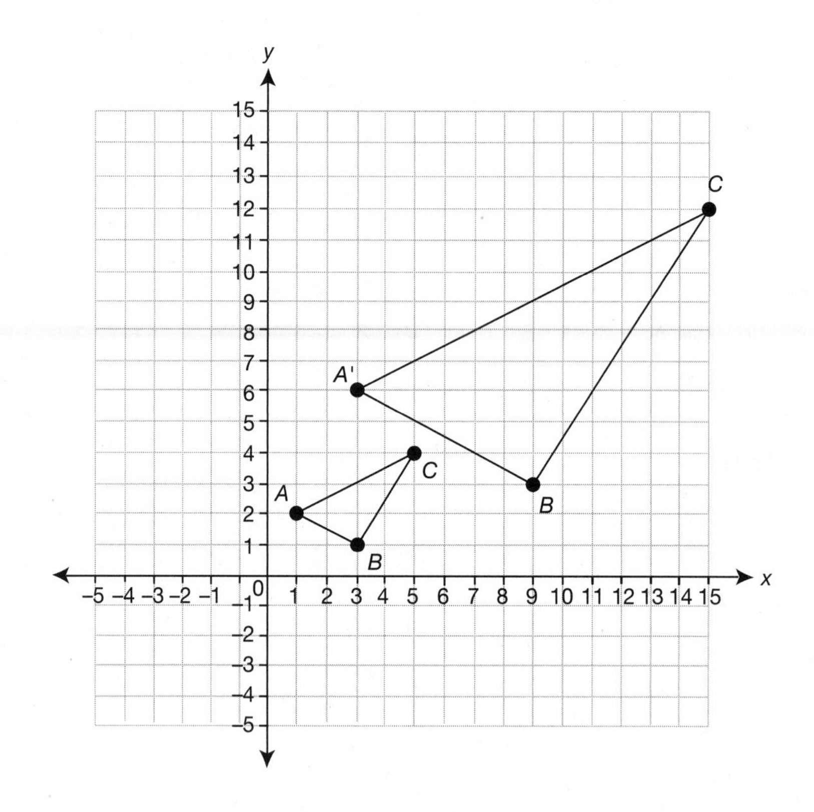

3. Here is one example of a dilation by a fractional factor, a factor of $\frac{3}{4}$:

Original Point	Image Point
D(1,4)	*D*'(0.75,3)
E(5,2)	*E*'(3.75,1.5)
F(4,−2)	*F*'(3,−1.5)

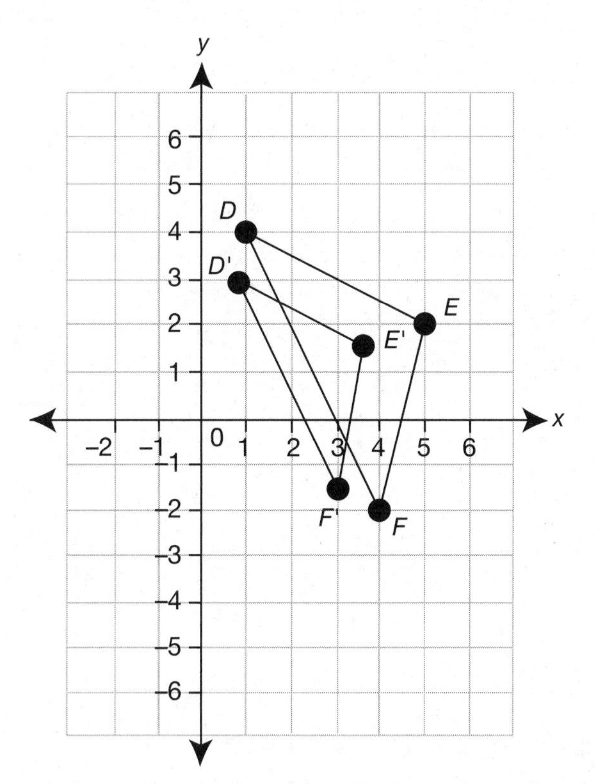

In this case, the image triangle is slightly smaller than the original. This is also a good example of how the images under a dilation will often overlap with the original figures.

Measuring Circles

In Chapter 5, you learned how to find the area of any polygon. But not all shapes are polygons, and sometimes your child will need to be able to find the lengths and areas of a circle or part of a circle. Understanding the relationships between the different parts of a circle is critical for your child's future performance in high school geometry and beyond. But not only that—it's also lots of fun.

Circles are such a deep part of nature that they appear nearly everywhere you look. Everything from infinitesimal atoms to the universe itself has this shape in common, which is why circles are so appealing and so useful. When students first learn about circles, there is an opportunity to make some very significant connections to life, the world, and the things that surround us every day, which are all related through a few simple ideas.

Parts of a Circle: Radius, Diameter, and Circumference

There are four different ways a circle can be measured: by its radius, diameter, circumference, and area. Radius, diameter, and circumference are all one-dimensional—or linear—measures, which means they can be measured in units like inches, feet, meters, etc.

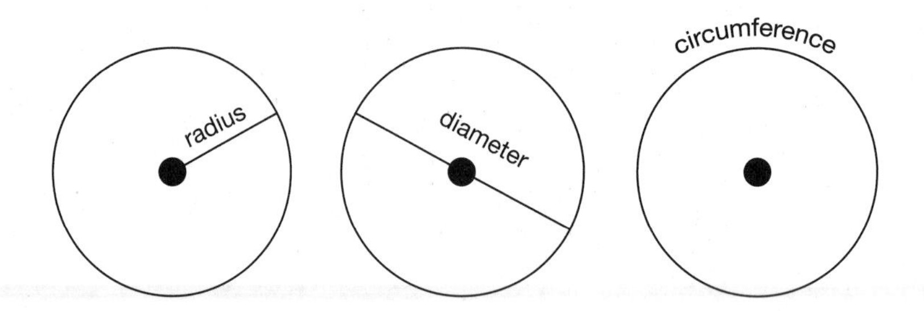

As you can see in the diagram, the **radius** of a circle is a line segment connecting the center of the circle to the outside of the circle. A radius can point in any direction, connecting the center to any point on the circle. In any given circle, all radii have the same length. If you want a tangible example of what radius means, try drawing a circle. It is nearly impossible to draw a perfect circle freehand. But with the help of a simple tool, it is not difficult. Try using a paperclip and two pencils. Hold one end of the paperclip in place with one of the pencils, and with the other pencil, draw the circle by rotating around the center. If you do this carefully (it takes some dexterity!), you will have a very nearly perfect circle, with a paperclip for its radius.

The **diameter** of a circle is any line segment that passes all the way across the circle and through the center. If you are standing on a circular track and your child is standing directly across the track as far away from you as he can be, the line segment between you and your child would be a diameter of the track.

The **circumference** of a circle is the length around the outside of the circle. Think of the circumference as the distance you would walk if you made one lap around that circular track, ending at your starting point.

Notice that the diameter of a circle is made up of two radii pointing in opposite directions from the center of the circle. This makes it simple to calculate the length of a diameter (abbreviated *D*) if you know the length of a radius (abbreviated *R*), or vice-versa. We have two formulas:

$$D = 2R \quad \text{and} \quad R = \frac{D}{2}$$

DIFFERENT FORMS OF THE FORMULA

Sometimes you may see the second formula expressed in different ways:

$$R = D \div 2,\ R = \frac{1}{2}D, \text{ or } R = 0.5D$$

All of these are correct. The previous form, however, which incorporates *D* into the numerator of the fraction, is a great setup for students' future work with algebraic fractions. Many students get to algebra without understanding that division and fractions really mean the same thing, and this is one way to reinforce that idea early on.

Examples

1. How long is the diameter of the circle below?

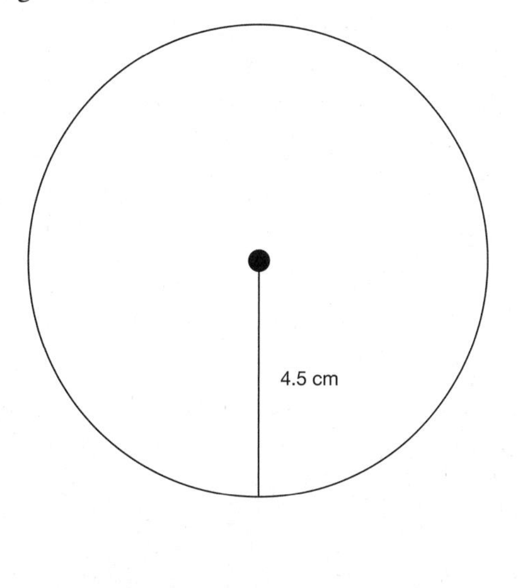

Since we know the length of the radius, we can double it to find the diameter. The radius is 4.5 centimeters, so 4.5 cm × 2 = 9 cm. The diameter is 9 centimeters.

2. How long is the radius of a circle whose diameter measures 12.75 feet?

Even without a diagram, we can use the formula to find the radius. However, you should encourage your child to always draw a diagram for any geometry problem, even when it is not given, because visualization is such an important part of developing geometric concepts. It is also a good habit to get into as soon as possible, as it will help justify solutions to more complicated problems later on.

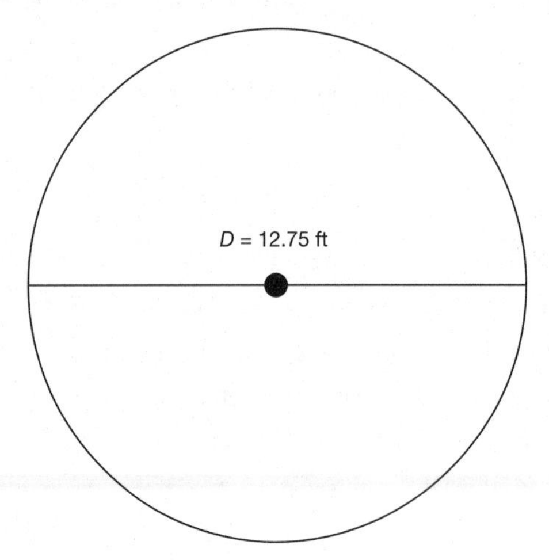

Since the diameter is 12.75 feet, we can find the radius by dividing 12.75 ÷ 2, which is 6.375 feet.

Circumference and Diameter

The relationship between circumference and diameter is one of the simplest and at the same time most mysterious ideas in all of mathematics. In the ancient world, this relationship was discovered by several different civilizations, and many mathematicians throughout history have attributed this relationship to mystical or spiritual origins.

If you have a piece of string, a ruler, and any circular object, you can discover the relationship for yourself—a nice activity to do with your child. Take a can of soup (or a bicycle wheel, or any other circle you find around the house). Measure the diameter, through the center of the circle.

Then wrap a string around the circumference of the circle. Make it tight, so that the string is stretched fully. If you measure the length of the string that went around the object, you will find that it is a little bit more than 3 times the diameter. The same relationship will be true, no matter what circular object you use—*the circumference is always a little more than 3 times the diameter.*

Of course, mathematicians are not often satisfied with descriptions like "a little more than." They want precision. So in time, mathematicians refined their estimate of the ratio of circumference to diameter. For any circle,

$$\frac{C}{D} \approx 3.14159265 \ldots$$

The digits of this ratio go on forever without ever terminating or repeating. In Chapter 3, we discussed two different kinds of decimals: terminating decimals, which stop after a certain number of digits, and repeating decimals, which repeat the same digits infinitely. But this number is a non-terminating, non-repeating decimal—a decimal that cannot be expressed in fraction form. For your child, this will be the first example of a new kind of number—an **irrational** number. We saw irrational numbers earlier in this chapter when solving Pythagorean theorem problems. Sometimes evaluating a square root will give an irrational result.

To make it easier to write this number, since the digits go on forever with no pattern, mathematicians have given it a name: π. π is the Greek letter pi, not to be confused with pie, which is pronounced the same way. π represents the ratio of a circle's circumference to its diameter. It is the same for any circle.

Calculating with π

Since the ratio π relates to any circle, no matter how large or small, we have some useful formulas:

1. $C = \pi D$
2. $\pi = \frac{C}{D}$
3. $D = \frac{C}{\pi}$
4. $C = 2\pi R$

Equations 1, 2, and 3 are all part of the same fact family, simply relating the same three numbers. Equation 4 is equivalent to Equation 1, but with $2R$ in place of D since they are equivalent. It is worth taking some time here to make sure your child understands how these four equations are related—it will pay off in the long run because these equations can be generalized to many common equations in algebra.

We can use these formulas to calculate circumference, radius, or diameter, given any of the other three.

APPROXIMATING π

In order to find the circumference of a circle, given the diameter, we need to multiply the diameter by π. But it is not convenient to multiply by a number whose digits go on forever with no pattern, so we need to approximate. There are two very common ways to approximate π:

In decimal form, most teachers encourage students to round π to 3.14.

In fraction form, you can use the fraction $\frac{22}{7}$, or $3\frac{1}{7}$.

If you convert $\frac{1}{7}$ to a decimal by long division like we did in Chapter 3, it comes out to $0.\overline{142857}$. It is not the fact that this decimal goes on forever that makes it a good approximation for π, it is the fact that the first few digits are so close to the first few digits of π: 3 and 14 hundredths.

MISCONCEPTION ALERT!

Many students make the mistake of thinking that $\frac{22}{7}$, 3.14, and π are equivalent. Emphasize the point for your child that $\frac{22}{7}$ and 3.14 are useful because they are *close* to π, but they are not equal to π.

Examples

1. What is the circumference of the circle below?

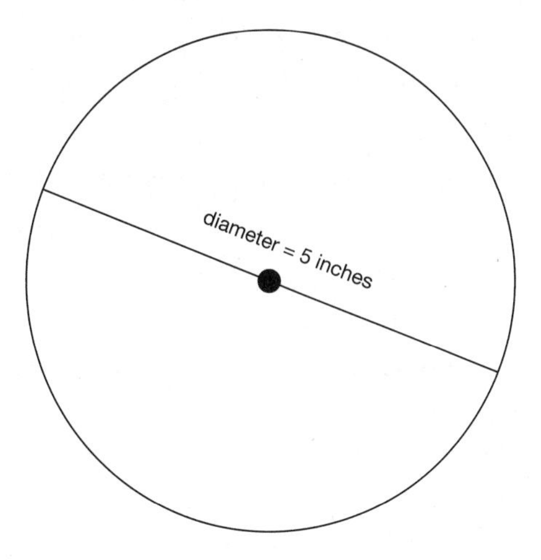

The diameter is 5 inches, so we need to multiply $5 \times \pi$. For this problem, let's use the decimal approximation 3.14: $5 \times 3.14 = 15.7$, so the circumference of the circle is approximately 15.7 inches.

2. What is the circumference of a circle whose radius is 12 feet?

Since the diameter is 2 times the radius, the diameter of the circle is 24 feet. We can multiply this by $\frac{22}{7}$ to approximate the circumference in fraction form:

$$24 \times \frac{22}{7} = \frac{528}{7} = 75\frac{3}{7}$$

So the circumference is approximately $75\frac{3}{7}$, or 75.43 feet.

3. What is the diameter of a circle whose circumference is 20 meters?

In this case, we could use the equation $D = \frac{C}{\pi}$ since we are trying to find the diameter from the circumference. Divide: $20 \div 3.14 \approx 6.37$ meters.

RELATING THE FORMULAS TO LIFE

You may remember learning the formulas for circular relationships when you were in school. Your child will likely have the advantage of a teacher who will present the formulas not as something boring to memorize, but as a way of summarizing what your child can see in the world around him.

You can help your child solidify this understanding of circles by asking him about the parts of any circle you encounter. What's the radius of a bicycle wheel? It's one of the spokes. What's the circumference of a tire? The treads are. Which part of a Ferris wheel do people ride on? The circumference. Which part lights up at night? The diameter and the radii.

Area of a Circle

In Chapter 5, we learned that the area of any polygon may be found by cutting the shape into rectangles and triangles. Circles are not considered polygons because they are not made up of straight line segments. It would be impossible to precisely count the number of square units in the following circle, or to fit the irregular pieces together to make square units.

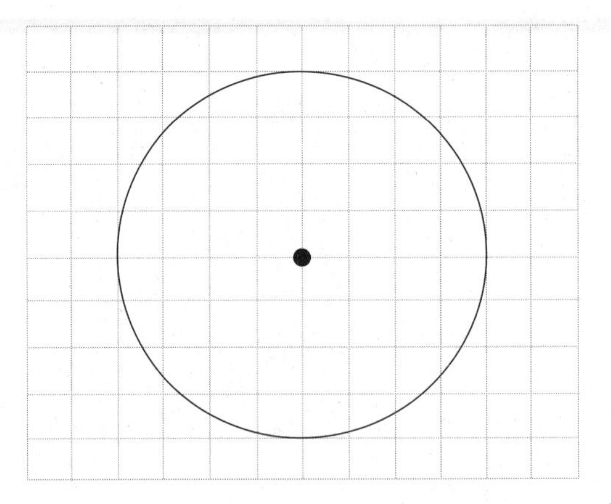

Since we cannot use the usual method to find the area of a circle, we need another method. Fortunately, there is a simple formula, which you may recall. There are many ways to come up with the formula for the area of a circle, which your child will likely see in higher mathematics. At the middle school level, your child should understand the following simple, informal way of deriving the formula.

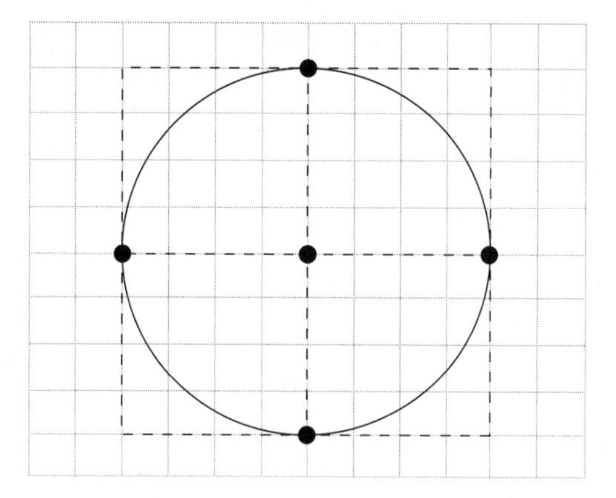

In this diagram we have quartered the circle and have covered each quarter with a square whose side lengths are the radius of the circle. So the area of each square is the circle's radius, squared, r^2. The area of the circle is clearly less than the total area of the 4 squares. If you count boxes, rounding partial boxes when necessary, you can see that the area of the circle is pretty close to but a little more than the area of 3 of the squares. What number do we know that is a little more than 3 and is hard to estimate?

The area of a circle is given by:

$$A = \pi r^2$$

If you know the length of the radius of a circle, you can find the area of that circle by squaring the radius and multiplying by π.

Examples

1. Find the area of a circle with a radius of 5 inches.

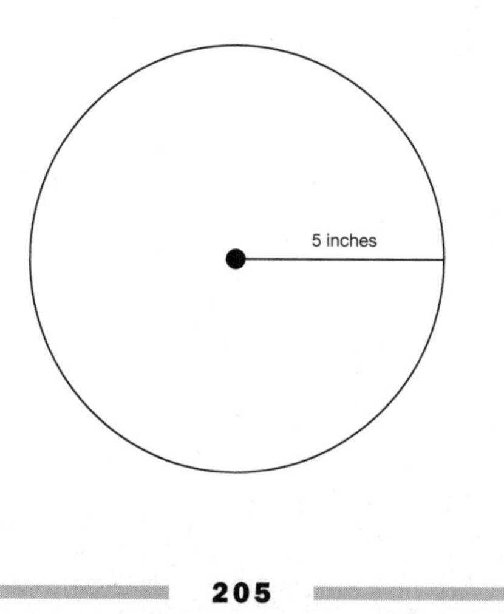

We plug 5 inches into the formula:

$$A = \pi r^2$$

$$A = \pi(5 \text{ in.})^2$$

$$A = \pi \times 25 \text{ in.}^2$$

$$A \approx 78.5 \text{ in.}^2$$

Notice the steps we followed here. Circle formulas are a great way for your child to practice good habits for the algebra ahead. By first showing the original formula, I am letting you know how I am going to solve the problem. I then make the substitution, and solve the problem one operation at a time.

2. Find the area of a circle with a diameter of 20 meters.

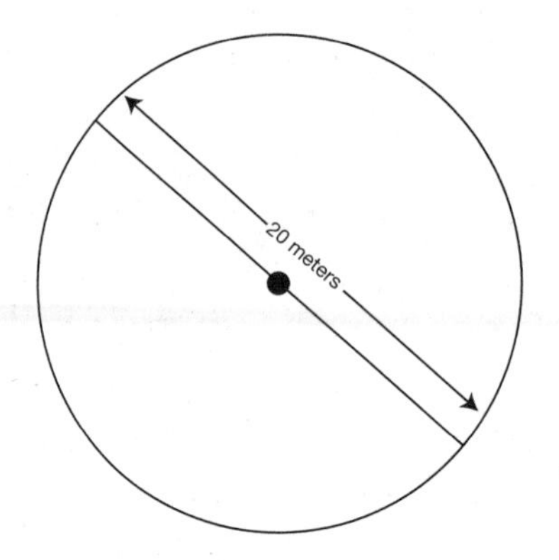

In this case, we must first find the length of a radius. Since the diameter is 20 meters, the radius is half that, or 10 meters. Now we can use the formula:

$$A = \pi r^2$$

$$A = \pi(10 \text{ m})^2$$

$$A = \pi \times 100 \text{ m}^2$$

$$A \approx 314 \text{ m}^2$$

Summary

In this chapter, we covered some of the most engaging topics in middle school math: angle relationships, the Pythagorean theorem, transformations, and circles. Many students who have disengaged from math class by middle school find themselves interested again when these topics come up because the geometry is concrete, visual, and challenging.

As Dr. Seuss put it in *The Cat in the Hat*, "it is fun to have fun, but you have to know how." If your child has fun doing math, then he will be willing to put in the extra practice to become more successful. That success will help your child continue to enjoy the subject. If your child has never enjoyed math before, or not recently, geometry is a great way to change that. When you catch your child having fun doing math, point it out and make a big deal of it—reward that positive attitude. For many middle schoolers, it is difficult to give oneself permission to enjoy math, for fear of how it might look to other students. Let your child know that it is O.K. to like math. Once he opens up to the idea, great things will follow.

In Chapter 7, we will shift our attention to algebra—the most dreaded topic for many students. As we cover the basic language of algebra—variables, expressions, and equations—you will learn a simple but effective questioning strategy that any parent can use to help a child with math, regardless of his or her own math knowledge.

The Two Most Important Words in Mathematics

How Why *and* Because *Can Unlock the Language of Algebra*

The Right Parent

If your family is like many I have encountered, then helping your child with math has probably become a struggle. Maybe it used to be easy (even fun!) back in elementary school. But by middle school, the increased rigor of the content—which may be approaching or even beyond your own math knowledge—and the new methods teachers are using, combined with the emotional challenges of adolescence, can make it extremely difficult to have a peaceful parent/child math session.

In many families, there is a natural choice for which parent (or uncle, cousin, grandparent, or neighbor) should be the math helper. Usually there is somebody in the family, most often a parent, who is viewed as the family's "math person." The choice of whom to go to for help is usually obvious to the child, if the child even has a choice.

But sometimes, even with the obvious choice of parent helping, homework time can be explosive. The default parent tries to help, which ends up leading to conflict, frustration, and unkind words. The non-math parent or other family member tries hard to stay out of it, only getting involved to clean up the mess after the sparks fly.

THAT GUILTY FEELING

When things go wrong, it is easy to feel guilty, like it is somehow your fault as a parent. I assure you, it is NOT your fault. Whatever you are doing to help your child with math, you and I understand that you are doing it with the best of intentions—with love and a hope for your child's future success at heart. Your child may not understand that (and at times you may lose sight of it yourself), but that is the reason why you try so hard to help. Nobody could ever fault you for anything that may go wrong—and you shouldn't blame yourself, either. You are doing exactly what you should do—you are doing your best, and you are doing what is best for your child. As a parent, sometimes nobody else but you can possibly understand what that means.

Over the years, I have observed families and discussed this struggle with many students and their parents. I have come to a simple, but counterintuitive conclusion: In most families, the parent trying to help the child is the *wrong parent*! Remember the story of my mother offering to help me with my calculus homework (it's in the Introduction)? While I would not describe my mother as mathphobic, let's just say math isn't her strongest subject, which is why I found the idea so hilarious at the time. To illustrate, I can tell you that 10 years after the calculus incident, she needed to learn how to multiply decimals for a work-related test, and after I showed her the lattice method, which your child may be able to explain to you, she said, "Wow, I never really understood multiplication before!" But what I realize now, much, much later, is that my mother could have been a huge help if she had only known a simple strategy: **Ask *why*, then ask *why* again.**

At the time, I was trying to make sense of the Fundamental Theorem of Calculus, one of the most beautiful and significant equations ever devised:

$$\int_0^x f'(t)\,dt = f(x)$$

Of course, my mother was completely unprepared to understand this equation. And I was completely unprepared to explain it, given that I was struggling to understand it myself. But if she had begun to ask *why*, she would have forced me to explain the parts of it that I *did* understand. If she didn't understand my explanation, she could ask *why* again and again, probing me until I could make sense of what the equation was trying to tell me. If we had to get down to the point where I was explaining why $1 + 1 = 2$, we would have gotten somewhere, because all along the way, I would have been making sense of the mathematics for myself.

I have often told families that they have the wrong parent trying to help, much to their surprise, anger, and frustration. What very few people understand is that good math help is more about listening to the student than talking to the student.

By asking your child questions about the math and listening to her understanding, you can come up with questions that will help her keep making sense of the math. The more you ask *why*, rather than explaining why, the more you will allow your child to tell you what she needs. Many of the tensions between children and their math-confident parents arise from the parent's desire for the child to master a specific method that the parent remembers, rather than building on what the child already understands.

If math help in your home has become a nightly battle, it may be time to make a substitution on the field. Let the other parent or another adult take a swing, even if that other person isn't much of a math person. Let that person seek his own understanding of the math, with your child's guidance, which will force your child to figure it out for herself.

A PARENT'S JOB

Ask *why*, then ask *why* again . . . and again, and again, and again, until you, your child, or both really understand it. *Why* is your most important tool, and *because* is your child's.

Introduction to Algebra

Algebra is the language of modern technology, so the importance of a strong foundation in algebra cannot be overstated. For many students, it is the most intimidating of subjects. For some, it is the end of the road. Students who are not successful in their first high school algebra course may be at high risk of not completing high school. In the 21st century, your child will be competing in an increasingly technical job market.

Algebra is an area where students often become frustrated and want to give up on math completely. As a parent, you need to remind your child that this is not meant to be easy, that the learning is in the struggle. I often tell my students that it took the greatest mathematicians in history thousands of years to come up with this stuff, and I'm asking them to learn it in just a few months. Patience, practice, and a positive attitude will all pay off in the long run.

And remember—you don't need to understand a thing your child is talking about to be able to help.

The Basics of Algebraic Language

If algebra is a language, then expressions are the words and phrases that make up that language. Every algebraic expression communicates something, and if your child can master their use, he will be able to interpret more complex ideas when they come along.

Throughout this chapter, it is important to remember that every symbol in math has a meaning. The more you can ask your child, "What does it mean?"—which is really a form of "Why?"—the more quickly your child will understand and become fluent in the language of algebra.

Constants and Variables

The basic building blocks of algebra are constants, variables, and operations.

A **constant** is a number that does not change. If you see 7 in an algebraic expression, it can only mean 7, nothing else.

A **variable** is a letter or other symbol, such as x, used to represent a number. A variable can represent a changing value, or a value that does not change but is unknown.

Examples

Here are a few algebraic expressions:

- $3x$
- $12 - n$
- $4(9p + 17)$

In these examples, x, n, and p are all variables. Their values are unknown, and may change or stay the same. 3, 12, 4, 9, and 17 are constants. No matter what happens to the variables, these numbers will always have the same values.

Addition and Subtraction

Addition and subtraction are simple to show in algebraic expressions. The expression $a + 5$ means exactly what it looks like—add 5 to a. If a is 10, then the expression has a value of 15. If a is 3, the expression has a value of 8. Subtraction works the same way, like $10 - b$. This can be really useful. If you are going shopping and you need to buy a $20 gift card in addition to other things on your list, then the total you will need is $20 + x$, with x standing for the total cost of your other items. If you are 20 years older than your child, then his age will always be $p - 20$, where p represents your age, which changes (sorry!).

Multiplication

Your child is probably used to using the symbol "×" to represent multiplication. In algebra, that symbol will seldom be used because it is easily confused with the most common variable, x. Instead, there are three different ways multiplication can be shown:

- With a dot, as in $4 \cdot 9 = 36$
- With a quantity before parentheses, as in $5(-6) = -30$
- With a constant next to a variable, as in $7x$, which means multiply x by 7. In this case, the number is called a **coefficient**.
- With two variables side by side. The expression ab represents the product of a times b.

If you need to buy 5 train tickets for your family but you don't know how much they will cost, you could let x stand for the cost of each ticket, and think of the total cost as $5x$. Here 5 is a coefficient that multiplies the variable x. Whatever the cost of one ticket is, the total cost is 5 times as much. When a variable has no coefficient, it is sometimes convenient to think of the coefficient as being 1. Since multiplying by 1 does not change a number, a variable with no coefficient is equivalent to a variable with a coefficient of 1. That is, $x = 1x$.

MISCONCEPTION ALERT!

Students will often make the mistake of saying "there's no number in front of the *x*, so *x* is 1." Listen carefully for this—if your child says it, help him understand that *x* isn't 1, but the *coefficient of x* is 1—that is, *x* is only being multiplied by 1.

Division

Like multiplication, division often looks different in algebra. Your child will seldom see the "÷" symbol used. Instead, division is usually expressed in the form of a fraction.

- $\frac{100}{n}$ means the same as $100 \div n$, whatever the value of n is.
- In the expression $\frac{4x}{y}$, x is multiplied by 4 and the product is divided by y.

Division may also be written with a slash, which is another way of representing a fraction. For example:

- $x/2$ means the same as $x \div 2$.

Terms

When an expression involves addition and/or subtraction, each of the values being added or subtracted is called a **term**. In the example $a + 5$, a and 5 are both terms. 5 is called a **constant term**, and a is a **variable term**. Any term that contains a variable is considered a variable term because it can change, even if there are constants in the term as well.

In the expression $5x - y + 4xyz$, there are three terms. It does not matter how complicated a term is, so $4xyz$ is still one term, even though it has a constant and three variables.

Writing Expressions from Words

Now that we understand some of the building blocks of algebraic language, let's begin using variables and expressions to represent known and unknown quantities.

Examples

1. Tyshawn and Angelina are comparing their ages. Tyshawn is 3 years older than Angelina. If T represents Tyshawn's age, then Angelina's age could be described as $T - 3$.

2. Suppose Liam has twice as many baseball cards as Dmitry. If D represents Dmitry's baseball cards, then we can express Liam's baseball cards as $2D$.

3. In Chapter 5, we worked with the area of a triangle. If you know the base length b and the height h of a triangle, the area is given by the expression $\frac{1}{2}bh$.

To get a little more abstract, we can also write expressions based on simple mathematical phrases, using what we know about the four operations.

Examples

1. The sum of 9 and x
   Since *sum* indicates addition, the expression is $9 + x$ or $x + 9$. In addition, order does not matter.

2. Half of a number
   In this case, we are not given a variable to work with, so we are permitted to choose our own variable. $\frac{1}{2}x$ will do, or $\frac{v}{2}$, or $0.5k$.

3. 5 less than p
Less than indicates subtraction. In subtraction, order does matter, so it makes a difference whether we write $5 - p$ or $p - 5$. 5 less than p means we are starting with p and decreasing by 5, so $p - 5$ is correct.

This example is a great place to practice asking your child *why*. Whether she comes up with $5 - p$ or $p - 5$, asking *why* that represents "5 less than p" will force your child to connect the algebraic expression with the word phrase. If she is having trouble explaining it, make up a scenario. You might have a conversation like this:

You: *If I have 20 dollars, and you have 5 less than me, how much do you have?*
Child: *15.*
You: *Why is it 15?*
Child: *Because I have 5 less, so I did 20 – 5.*
You: *So if I have p dollars and you have 5 less than me, how much do you have?*
Child: $p - 5$.

Some expressions will call for two or more operations written together. Before we can write expressions like these, we need some experience with the order of operations.

Order of Operations (and the Problem with Dear Aunt Sally)

As your child gets into more sophisticated mathematics, she will often be faced with problems where several operations are being indicated at once. For example:

- $3 \times 2 + 9$
- $-4(5 - 8)$
- $15 - \frac{1}{2}(4 + 2)^2$

In order to make sure that everybody who sees expressions like these evaluates them in the same way, mathematicians long ago developed an order in which operations should be performed, regardless of how they are written.

Do you remember the old mnemonic device "Please Excuse My Dear Aunt Sally"? For many years, this has been used to help students remember the correct order of operations:

Parentheses
Exponents
Multiplication
Division
Addition
Subtraction

Unfortunately, for as long as dear Aunt Sally—or **PEMDAS** for short—has been around, students have been making major errors when using the order of operations.

Let's look at a classic example of the problem with PEMDAS.

$$12 - 4 + 7$$

When I give a problem like this to my 8th-grade students for the first time, the class ends up divided pretty evenly between two answers: 1 and 15. The students who get 1 reason that PEMDAS prescribes addition before subtraction, so they first add 4 and 7 to get 11, then subtract 12 – 11.

This is incorrect. Many teachers teach students that addition and subtraction should be done in the same step, left to right, but this is a rule that does not seem to stick. Instead, let's remember how we were thinking of subtraction in Chapter 1. We avoided it completely by turning all subtraction signs into addition by the opposite number. In this case, we can change the expression from before:

$$12 - 4 + 7$$

$$12 + (-4) + 7$$

Now we have three numbers to add together, and as we will see shortly, it does not matter what order we add them in: We always get a result of 15.

The previous discussion applies to multiplication and division as well as addition and subtraction. For example:

$$24 \div 3 \times 2$$

Many students interpret PEMDAS too literally and multiply before dividing, ending up with $24 \div 6 = 4$. But the original expression is the same as

$$24 \times \tfrac{1}{3} \times 2,$$

which is 16.

A BETTER ORDER OF OPERATIONS

Instead of PEMDAS, try eliminating subtraction and division first.

Rewrite $32 \div 4$ as $32 \times \frac{1}{4}$.

Rewrite $-25 - 8$ as $-25 + (-8)$.

Once you've gotten rid of those pesky inverse operations, you can follow the usual order of operations: Parentheses, Exponents, Multiplication, and Addition.

Parentheses

Parentheses, also known as **grouping symbols**, are a wonderful invention. They allow mathematicians to change the order of operations as needed. Say, for example, that you want to multiply 5 by the sum of 14 and 3. If you simply write $5 \times 14 + 3$, the order of operations would connect the 5 and the 14 first, so you would add 3 to 70, rather than multiplying 17 by 5. By using parentheses, $5(14 + 3)$, you can communicate to any other mathematician that the sum is to be found before multiplying.

In short, parentheses tell you to **think of everything inside as a single number**. My mother taught me that!

A fraction bar can also be used as a grouping symbol. For instance, in the expression $\frac{2+9}{5} - 16$, addition of 2 and 9 is the first step because these numbers are grouped together in the numerator.

Examples

Let's use our new and improved order of operations to simplify a few expressions.

1. $3(5 - 1 + 8)$

To avoid the trouble that comes when students do addition before subtraction, start by changing the subtraction to addition. Then you can add the 3 numbers in parentheses, in any order, and finally multiply the sum by 3:

$$3(5 + (-1) + 8)$$

$$3(12)$$

$$36$$

2. $\frac{20 \div 5}{2} + 1$

Again, a fraction bar acts as a grouping symbol, just like parentheses. Everything in the numerator is treated as a single number, and the same goes for the denominator. So first, we have to simplify the numerator. In this case, we do not need to convert the division in the numerator to multiplication. Remember that the point of that conversion is to avoid confusion when multiplication and division are written together. So we simplify the numerator, divide by 2, and finally add the 1:

$$\frac{20 \div 5}{2} + 1$$

$$\frac{4}{2} + 1$$

$$2 + 1$$

$$3$$

3. $(2 - 7)^2 + (6 + 4)^2$

You may recognize this as an example of the Pythagorean theorem. The parentheses are there to tell us to perform $2 - 7$ and $6 + 4$ before doing anything else, treating $2 - 7$ and $6 + 4$ as single numbers. Once we've done that, we can evaluate the exponents and add:

$$(2 - 7)^2 + (6 + 4)^2$$

$$(-5)^2 + (10)^2$$

$$25 + 100$$

$$125$$

Writing Expressions with Multiple Operations

1. 8 more than the product of 7 and j

"8 more" indicates addition. The 8 is being added to a product—the product of 7 and j. So first, 7 and j are multiplied together, then 8 is added. The expression could be $7j + 8$ or $8 + 7j$.

2. 16 times the sum of 90 and k

In this case, we are multiplying 16 times a sum. The first thing to be done is find the sum. The order of operations usually has us perform multiplication before addition, so to change the order, we need to use parentheses. First we need to find $(90 + k)$, then multiply it by 16. The final expression is $16(90 + k)$.

You could ask your child to explain why $16(90 + k)$ represents "16 times the sum of 90 and k." The conversation might go something like this:

You: *Why does that represent 16 times the sum of 90 and* k*?*
Child: *Because it does.*
You: *But where did it come from?*
Child: *From the words. The parentheses mean times.*
You: *So why are the 90 and* k *in the parentheses?*
Child: *Because you have to add them first.*
You: *So why can't you just write 90 +* k *× 16?*
Child: *Because multiplication comes first.*

You get the idea. You may read this dialogue and feel that you don't have enough understanding to think of these questions on your own. That's fine. There really aren't any *correct* questions to ask—the point is to just *keep asking* until you or your child or both get it. The more you ask about the meanings and reasons for things, the better your questions will be. By asking *why*, you will be encouraging her to organize what she has learned, putting the information in an order that makes sense. Try asking your child to explain the expressions for the next few examples.

3. 8 less than the product of 9 and u

Expression: $9u - 8$

4. Half of the sum of 5 and an unknown number

Expression: $\frac{x + 5}{2}$
(For this one, any variable will do—it doesn't need to be x).

5. 4 times 11 less than m.

Expression: $4(m - 11)$

Equivalent Expressions

Two expressions are equivalent if they always have the same value. For example, the expressions $3 + 2$ and $4 + 1$ are equivalent because both sums are equal to 5. When two expressions are equivalent, we can make an equation like $3 + 2 = 4 + 1$.

When variables are involved, two expressions are equivalent only if they are *always* equal, no matter what values the variables may take on.

Examples

1. x is equivalent to $x + 0$ because no matter what value x has, adding 0 to it does not change its value. So $x = x + 0$.

2. a is equivalent to $1a$. Multiplying any number by 1 does not change the value of the number. So $1a = a$.

The Commutative Property

The **Commutative Property** states that in addition and multiplication, the order of the numbers does not matter. For example, $4 + 7$ is the same as $7 + 4$, and 3×2 is the same as 2×3. To generalize, we can say that for any numbers a and b:

$$a + b = b + a$$

$$ab = ba$$

One reason why I teach my students to avoid subtraction is because it is not commutative. $7 - 3$ is 4, but $3 - 7$ is -4. Division is also not commutative: $12 \div 3 = 4$, but $3 \div 12 = \frac{1}{4}$.

The Associative Property

The **Associative Property** states that when adding or multiplying three numbers, it does not matter which two you operate on first. For example, when adding $5 + 3 + 13$, it makes no difference whether you add $5 + 3$ first and add the sum to 13, or you add $3 + 13$ first and add the sum to 5.

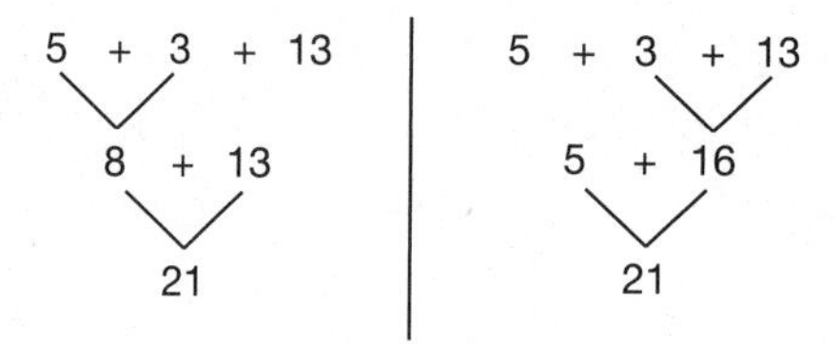

The same is true of multiplication:

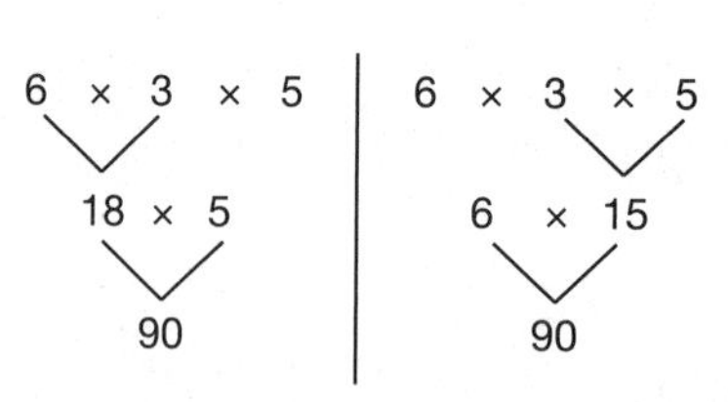

The Associative Property can also be written in variable form to generalize. For any numbers a, b, and c:

$$(a + b) + c = a + (b + c)$$

$$(ab)c = a(bc)$$

In other words, any time we need to add or multiply a bunch of numbers together, we don't need parentheses to group the numbers, since order does not change the sum or product.

Like the Commutative Property, the Associative Property does not work with subtraction or division, which has a lot to do with the problem with PEMDAS. Earlier we worked with the example $12 - 4 + 7$. If you think of the 4 as being positive, then it matters very much which order you follow in adding or subtracting the terms:

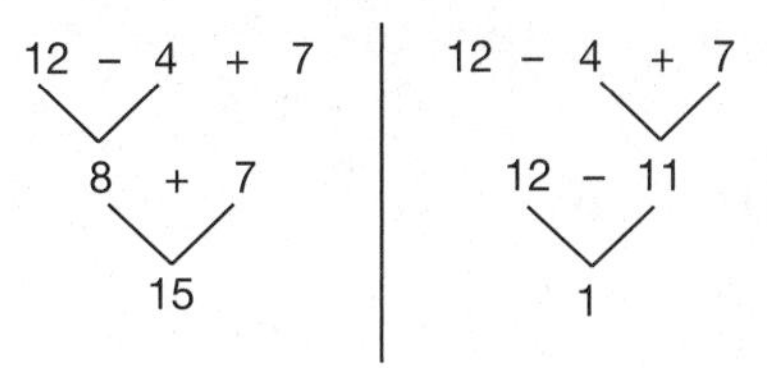

However, if you convert the subtraction to addition, then you will only be dealing with addition, and the Associative Property holds true:

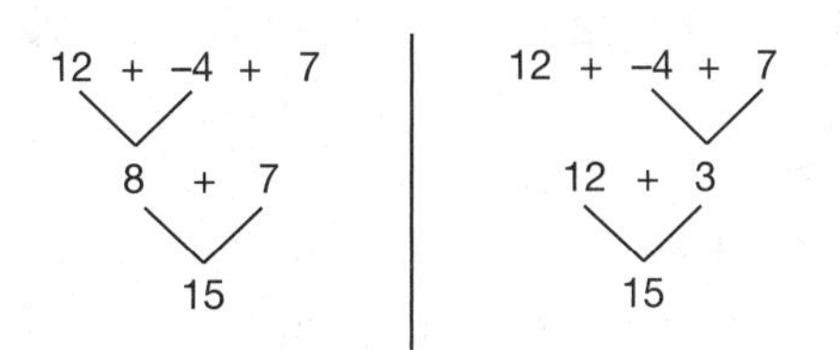

Combining Like Terms

One way to generate two equivalent expressions is by combining like terms. Recall that **terms** are parts of an expression that are being added or subtracted with each other. An expression may contain many different kinds of terms, and it is important to recognize like terms. **Like terms** are terms that have the same variable (or no variable).

Examples

1. x and $3x$ are like terms because they have the same variable.

2. $4k$ and $7j$ are *not* like terms because they have different variables.

3. The expression $2k + 5p - 3k + 6$ has two like terms: $2k$ and $-3k$. Since $5p$ has a different variable and 6 has no variable, they are unlike terms.

4. All of the terms in the expression $28 + 16.\overline{3} + -4 + \frac{7}{8}$ are like terms because they are all constants.

   Combining like terms means to add or subtract them in an expression. To use the examples above, since x and $3x$ are like terms, they may be added together: $x + 3x = 4x$. Remember that x is equivalent to $1x$. So to combine like terms, we simply add their coefficients. This works because $3x$ is the same as $x + x + x$. So $x + 3x = x + x + x + x = 4x$.

Examples

5. $7p + 5p$ is equivalent to $12p$: $7p + 5p = 12p$

6. $14s - 9t + 10s$ is equivalent to $24s - 9t$. Since s and t are different variables, their terms cannot be combined: $14s - 9t + 10s = 24s - 9t$

A GREAT TIME TO ASK *WHY*

Try asking your child why unlike terms cannot be combined. For instance, in #6 above, why can't $24s - 9t$ be simplified to 15? If your child answers "Because they're not like terms," or "Because they're different," keep pressing with more *why* questions—try to get to the root of it. Later, when your child sees expressions like $9xy^3 - 12x^2y$, it will be helpful to have that deep understanding of what like terms are.

The Distributive Property

The **Distributive Property** is a great tool for mental math, as well as one of the most widely used techniques in all of algebra. The Distributive Property states that multiplying a factor by the sum of two other numbers is equivalent to multiplying that factor by both of the addends, and then combining.

Let's see an example. On the left, we will follow the order of operations, adding what is inside the parentheses first. On the right, we will apply the Distributive Property.

$4(3 + 6)$	$4(3 + 6)$
$4(9)$	$4 \cdot 3 + 4 \cdot 6$
36	$12 + 24$
	36

Both ways give us the same value of 36. To put it simply, 3 groups of 4 and 6 groups of 4 is the same as 9 groups of 4.

The Distributive Property also works with subtraction (since subtraction is just addition by the opposite). This helps with mental math. Suppose, for example, you need to multiply 8 by 19. You can rewrite 19 as 20 – 1 to make things easier:

$$8 \times 19$$

$$8(20 - 1)$$

$$8 \cdot 20 - 8 \cdot 1$$

$$160 - 8$$

$$152$$

When I show this method to students, they often are surprised that they have already been using the Distributive Property for a long time, without knowing it. Have you?

To generalize the Distributive Property in variable form, for any numbers a, b, and c:

$$a(b + c) = ab + ac$$

When it comes to algebraic expressions, your child will often see examples like $4(x + 5)$. The Distributive Property says that adding $x + 5$ and multiplying the sum by 4 is equivalent to finding the sum of $4x$ and 20.

There's another way to think about the Distributive Property. We have seen several times that multiplication is the same as repeated addition. So $4(x + 5)$ means we are adding $x + 5$ to itself 4 times.

$$(x + 5) + (x + 5) + (x + 5) + (x + 5)$$

The Associative Property tells us we do not need the parentheses, so we can rewrite this as

$$x + 5 + x + 5 + x + 5 + x + 5$$

The Commutative Property tells us we can switch things around in addition, so we can rewite this again:

$$x + x + x + x + 5 + 5 + 5 + 5$$

And now we can combine like terms:

$$4x + 20$$

This is one of those times where it's okay to say, "Math is fun!"

I also like to show it as a classic vertical addition problem. For example, to distribute $5(2a - 4b)$, you could write

$$\begin{array}{r} 2a - 4b \\ 2a - 4b \\ 2a - 4b \\ 2a - 4b \\ +\ 2a - 4b \\ \hline 10a - 20b \end{array}$$

FOIL: The Double Distributive Property

You may remember the mnemonic **FOIL** from your high school algebra classes. Sometimes your child will need to multiply two algebraic expressions together, where each expression has two or more terms, such as $(x + 2)(3x - 5)$. FOIL stands for "First, Outer, Inner, Last," meaning that you multiply the first term in each parentheses, then the outer terms, etc., and add them all together.

Of course, we have other methods that make more sense these days.

For one thing, we can think of this problem as an extension of the regular Distributive Property. If we think of the first set of parentheses as representing a single thing, we can distribute that thing over the two terms in the second parentheses:

$$(x + 2)(3x - 5) = (x + 2) \bullet 3x - (x + 2) \bullet 5$$

We can rearrange the right half of this equation using the Commutative Property:

$$3x(x + 2) - 5(x + 2)$$

And now we can distribute the $3x$ and the 5 into the parentheses:

$$3x^2 + 6x - 5x - 10$$

(Remember from our work on the Pythagorean theorem that $x \bullet x = x^2$.) And finally, we combine like terms: $6x - 5x = 1x$, or just x. So:

$$(x + 2)(3x - 5) = 3x^2 + x - 10$$

The expression on the right side of this equation cannot be simplified any more.

Another way to think of a problem like this, where two two-term expressions (called **binomials**) are multiplied, is by using an area model like we did in Chapter 5:

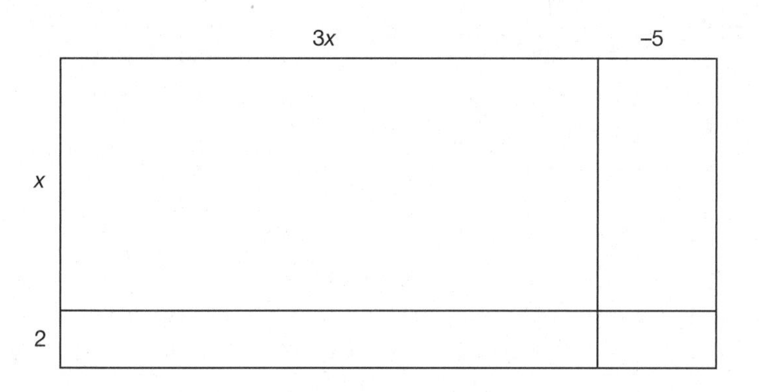

The left side of the rectangle represents $x + 2$ and the top represents $3x - 5$. It may be confusing to see a subtraction represented as an area here. How can you have a negative area? But we are just using the area model to help us figure out what to do with the numbers—it is not meant to represent anything real at this point.

Now we can find the area of each piece of the rectangle.

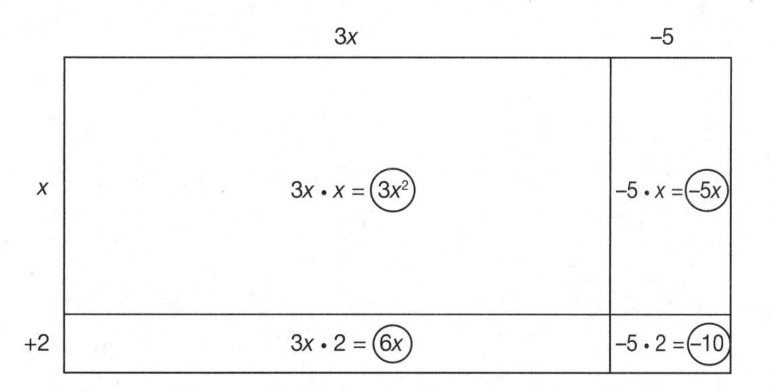

And just like before, when we add them up, the $6x$ and the $-5x$ combine to give us just x. Our total is:

$$3x^2 + 6x - 5x - 10 = 3x^2 + x - 10$$

Writing Algebraic Equations

Let's examine the difference between algebraic expressions and equations.

Expressions	Equations
y	$6x - 10 = 44$
$3x + 5$	$y = 4x + 9$
$\frac{29(a + 13b)}{41}$	$6a + 4b = 3x - 9a$

The examples on the left side of the table are all expressions. They may be as simple as just a variable (y), or they may give us a set of instructions for operations to do, like multiply x by 3, then add 5. If we continue to think of algebra as a language, expressions are phrases or sentence fragments, like "on your way home from school," "my favorite book," or "don't forget." They may give us something to do, but they do not convey a complete idea.

Equations, on the other hand, are complete sentences. The = sign acts like a verb in the sentence, telling us exactly what the relationship is between the numbers. The first example, $6x - 10 = 44$, tells us that if we multiply x by 6 and then subtract 10, the result will be 44. We can use this information to figure out what x is, and in Chapter 8, we will. The second example tells us that there is a specific relationship between the variables x and y. We do not know what value x has, but if we did, we could multiply that number by 4 and add 9 to find the value of y. We will work with equations like this in Chapter 9.

One essential skill that your child will depend on for as long as math is part of his life (that is, as long as he lives) is the ability to write algebraic equations given certain information.

The world is full of numerical relationships that affect us in our everyday life. Nothing ever happens in isolation, and whenever something changes, something else usually changes along with it. If you drive your car, your fuel gauge will go down. If you buy something, the amount of money you have will change. If your child studies, her grade will (hopefully) be affected. If you exercise, you will be healthier.

Each of these situations involves two variables. Remember that a variable in algebra can be used to stand for a number that changes. When two variables are involved, they may both change, and they may change in relation to one another. The more minutes you spend working out, the more calories you will burn. The more hours you spend babysitting, the more money you will earn.

When two changing quantities (like hours and money) have a specific mathematical relationship, it is often helpful to think of one variable as the **independent variable**, and the other as the **dependent variable**. The cost of refueling your car *depends* on how many gallons of gas you buy—"gallons" is the independent variable, and "cost" is the dependent variable.

Earlier in the chapter, we wrote expressions based on the first two examples below. Now, let's turn them into equations.

Examples

1. Tyshawn and Angelina are comparing their ages. Tyshawn is 3 years older than Angelina.

Let's use T to stand for Tyshawn's age, and A to stand for Angelina's. If Tyshawn is 3 years older, then we can add 3 to Angelina's age to find Tyshawn's age. Therefore,

$$T = A + 3$$

Many students have a misconception here. They reason that since Tyshawn is the older person, the +3 must go on the T side of the equation, and they write $T + 3 = A$. This equation says that 3 is added to Tyshawn's age to find Angelina's age.

If your child is writing basic equations, ask her to explain why she came up with a certain equation. Asking *Why are we adding* $A + 3$*?* or *Why are we adding* $T + 3$*?* might help your child make enough sense of the equations to be able to overcome that misconception or avoid it completely.

We can also reverse this idea, subtracting 3 from Tyshawn's age to find Angelina's age:

$$A = T - 3$$

2. Liam has twice as many baseball cards as Dmitry.

We can use D to represent the number of baseball cards Dmitry has, and L to represent Liam's number of baseball cards. Since

Liam has twice as many cards as Dmitry, we need to multiply Dmitry's amount by 2 to find Liam's amount:

$$L = 2D$$

Again, we can also do the reverse, dividing Liam's number of baseball cards by 2 to find out how much Dmitry has:

$$D = \frac{L}{2}$$

3. Suppose a cup of coffee at Alice's favorite coffee shop costs \$1.85. The number of cups that Alice buys in a week may vary, as will the total amount she spends. The total amount that Alice spends depends on how many cups of coffee she buys. Let's choose variables. Let n represent how many cups of coffee Alice buys and let c represent the total cost. We can write an equation: $c = 1.85n$. This handful of symbols says a lot. However many cups of coffee Alice gets, multiply it by 1.85 to find out how much she has spent in total.

4. Vanessa was born in 1997. Vanessa's age, v, can be expressed in terms of the current year, y:

$$v = y - 1997$$

Whatever year it is now, subtract 1997 to find the age Vanessa will turn this year.

5. On a road trip, the Jones family drives at an average speed of 68 miles per hour. Let d represent the distance they have traveled, and t the amount of time they have spent driving.

$$d = 68t$$

Some of these examples may seem like trivial ideas, hardly worth writing an equation for. If your child is starting to learn these concepts for the first time, she will encounter many such simple equations. These are important, but they are not the ultimate goal—we are using these simple ideas as practice for later on, when things will get complicated.

A great way to deepen your child's algebra sense is to ask *why* questions about the equations she is writing, without reading the context. Let your child explain to you what the equation represents, and why its symbols and operations work to show the idea she is trying to convey. She will have to explain what the variables stand for and explain what their relationship is in plain words so that she can make you see the connection between the words and the algebraic symbols.

Summary

The topics we've covered in this chapter are some of the most important in all of middle school math. Algebra is an essential language in the 21st century. Understanding how algebraic expressions and equations represent the relationships between numbers and variables in the world will allow your child to interpret all kinds of equations later on in algebra, geometry, and beyond. By asking your child *why* and pressing for more and more thoughtful *because* answers, you can help your child develop a deep understanding of algebraic language, without which he will face many struggles later on.

Always remember that your role as a parent is to help your child understand the math—not to force your own understanding on him. This can be very uncomfortable for some parents, but in the long run, it is much better for the child. That is why asking questions (rather than giving information) is often the best way to help your child with algebra and any other math topic.

In Chapter 9, we will continue our work with middle school algebra by learning how to solve equations. Our strategy will focus on how to interpret what your child means when he or she says, "I don't know."

The Many Meanings of *I Don't Know*

Learning to Turn Unknowns into Knowns

Three Dreaded Words

If *why* and *because* are the most important words in math, *I don't know* is the most dreaded, overused, misinterpreted, and dangerous phrase. For a parent, these words can be incredibly frustrating to hear, and the beginning of many a battle fought at kitchen tables over math homework.

The thing is, *I don't know* is often difficult to interpret. Many parents, after hours, days, weeks, or years of trying with increasing frustration to help their children with math, take these words as a personal attack, an act of defiance. In my classroom, we have a rule: "No IDKs." Students sometimes use *I don't know* to stop their learning, to get a teacher or a parent off their back when they feel vulnerable or uncertain. When a student in my class says, "I don't know," I offer alternatives that will continue the learning process, and I never let the student off the hook. This chapter will help you learn to do the same for your child.

There are many things your child might mean when she says, "I don't know," some of which may be things your child doesn't yet know *how* to say—so "I don't know" may be the only way to express them. Below you will find a list of the good meanings—those that represent and contribute to a positive attitude toward math—and the bad ones.

The Good Meanings

Most of the time, when your child says, "I don't know," he will really mean it. There are many ways you can respond. Here are some actual sentiments your child might be trying to convey, with ideas for what you can do to help:

When "I don't know" really means "I don't know."

Sometimes *I don't know* truly means *I don't know*—nothing more, nothing less. Faced with an equation to solve, say $9x + 5 = -22$, your child gets stuck and those words come out. As a parent, you might know the answer, and you might be tempted to give it to your child. *Don't.* This kind of *I don't know* is deeper than not knowing the answer to a particular problem. There is something about the problem that your child does not understand. Giving her the answer will not help her clarify her thinking; it will just help her get her homework done faster. Instead, ask questions to figure out what the problem is, as we discussed in Chapter 7. Does she understand what the equation means? What is happening between x and -22? Is the negative number throwing your child off? What if that were a positive number instead? By asking these kinds of questions, you can dig down to find the exact reason your child is confused and work with her on building that deeper understanding.

When "I don't know" means "I'm not sure yet—please help me figure it out."

This is what we hope our child will mean when he says, "I don't know," and in truth, it probably accounts for more of your child's *I don't knows* than you would think. Your child, like all students, truly wants to be successful and to understand what is being taught. What makes this meaning of *I don't know* so tricky to interpret is that it can be said with many different tones, in many different contexts. It is usually best to assume this is what your child means, until your child proves otherwise—innocent until proven guilty. In this way, you will always give your child the benefit of the doubt as to the meaning of *I don't know*, and continue to build on his positive attitude toward math.

When "I don't know" means "I don't understand your question."

This often happens when a parent who knows math pretty well is trying to help. The child has learned the concept a certain way from her teacher, and now the parent is asking it a slightly different way. This can be vexing and can lead to a big emotional reaction if not handled properly. A parent might be tempted to ask the question again, a little louder, maybe a little slower, but that doesn't help.

Instead, say to your child, "Tell me what you understand about this problem." As we discussed in Chapter 7, this is a good time to follow up with a few *why* questions. Instead of trying to make your child understand it *your* way, figure out how your child already understands it, and help her build on that understanding.

When "I don't know" means "Leave me alone; I can do it myself."

This might not seem like a positive meaning of *I don't know*. As a parent, you want to help—you want to make sure he understands it, and you want to feel connected. There's good news and bad news. Your child is growing up and maybe doesn't want you around that much any more. That can be a good thing—after all, isn't the point of parenting to raise independent, successful adults? If your child is confident in his math and doesn't seem to want your help, give him time. See how he does. Even though you may have been the parent who checked your child's homework every night in elementary school, middle school might be the time to let that go.

This doesn't mean you should switch to a totally hands-off style. Call the teacher and check in to see how your child is doing. The teacher can tell you whether your child's self-assessment is accurate. If it is, great. If not, then you can revise your plan. A conversation among the three of you—child, parent, and teacher—can clear the air and help you come to an agreement about how much help or supervision is necessary.

The Not-So-Good Meanings

As you've probably already discovered, sometimes *I don't know* means something much less benevolent. Middle schoolers are a hotheaded bunch in general. Even if your child has always been cool, calm, and collected, the challenges of the age—the hormones of adolescence, the social pressures of middle school, the physical changes, the increased academic rigor, and the discovery that the adults in your life are nothing but humans—will likely change that, at least temporarily. Below are some of the less pleasant ways your child might mean *I don't know*, and suggestions for how you might handle each situation.

When "I don't know" means "I'm just trying to get this homework done so I can do something else."

This is the ugly side of *leave me alone; I can do it myself.* Unlike that one, *I'm just trying to get this done* often comes with a bit of denial, meaning that your child may think she understands the math better than she actually does. In the moment, you may want to back off. Check in with the teacher, and if it sounds like things are not going well, take your child out for ice cream and have a conversation about how math is going. The sugar and the time with you will (hopefully) open her up to a genuine conversation. If you don't think it will, then talk to a favorite aunt or uncle or adult friend about doing the same to get you the information you need. Your child may try to hide things from you, but you are older, wiser, and more resourceful (and persistent!).

When "I don't know" means "I don't care. No, really, I don't care."

It's very common for students in middle school and high school to feel like they don't care about math any more. The teachers are not connecting math to their lives as much as they did in elementary school. The work is becoming much harder, and therefore your child is having more opportunities to struggle, or even fail. The work may seem purposeless, and with everything else going on in your child's life right now, it can be very easy to dismiss math as insignificant.

One fortunate thing about this *I don't know* is that it is very easy to recognize. If you're paying attention, you can see this coming long before it hits. Students don't stop caring overnight. The other fortunate thing is that these words are rarely true. Your child really does care—about math, about school, and about success (about you, too, though you might not hear it for the next few years). Your job as a parent at this point is to remind your child why she cares—not why you care, but why *she* cares. What does she want to do when she grows up? How is math important in her favorite hobbies, activities, games? What happened to the pride she used to take in doing well in math? This may be a time to use some of the attitude-improving strategies from Chapter 1.

When "I don't know" means "I give up."

This one is similar to *I don't care*, but more momentary. *I don't care* is the accumulation of a long series of frustrations. *I give up* is more immediate, as in *right now, I don't understand how to do this, and I'm not going to try*. This is a time when it's usually helpful to acknowledge your child's feelings and talk about them—briefly. Be cautious here, though, as your child may see an opportunity to use this conversation to get out of doing the math. In the end, you must always bring the discussion back to the math, and how important it is to persevere through this challenge. Help your child contextualize his frustration. This is a perfectly natural feeling to have, and it's appropriate for the situation. In fact, if he weren't frustrated at some point, it would probably mean he's not learning much because the math *shouldn't* be easy all the time. Teach your child how to channel that frustration into more useful emotions—determination, perseverance, and drive. Who are your child's heroes? Without a doubt, they are people who have overcome seemingly impossible challenges, which is what makes them heroes. Your child might not see the connection between a math problem and a sports championship or a hit song, but it's your job to help him see that hard work applies to all disciplines—including math.

Figuring Out the Unknowns

As we now turn our attention to the algebra problems your child will see in middle school math, we will keep thinking about times when your child might say, "I don't know," and what it might mean.

For many adults, the word *algebra* is synonymous with the phrase *solve for x*. This is an unfortunate side effect of all those hours of solving one-variable equations that most of us were subjected to in high school. Algebra is a much richer, more useful subject than that single topic. However, before we tackle all the things algebra can do, let's focus on solving for x.

Solutions to Equations

In Chapter 7, we discussed the different meanings of variables. x can be used to represent a number that is changing or fixed, known or unknown. When an equation is given, like $7x - 12 = 58$, x must have a specific value to make the equation true (it is fixed), but that value may be unknown.

A **solution** to an algebraic equation is the value of the variable that makes the equation true. For any equation, if you know (or think you know) a solution, you can check whether it is correct by substituting the solution for the variable and simplifying both sides of the equation with the order of operations.

For example, is 9 a solution to the equation $4x - 13 = 27$?

To find out, we substitute 9 for x:

$$4(9) - 13 \stackrel{?}{=} 27$$

Here we use the "$\stackrel{?}{=}$" symbol because we do not know yet whether it's correct to use an "=" sign—that is, we do not know whether both sides of the equation have the same value.

Now simplify by following the order of operations:

$$36 - 13 \stackrel{?}{=} 27$$

$$23 \neq 27$$

Since the two sides of the equation are not equal, 9 is not a solution to the equation.

Examples

1. 12 is a solution to the equation $3x = 36$ because replacing x with 12 makes the equation true:

$3 \times 12 = 36$

2. 70 is a solution to the equation $a - 21 = 49$ because replacing a with 70 makes the equation true:

$70 - 21 = 49$

3. 9 is *not* a solution to the equation $5x + 3 = 4x + 10$. If you substitute 9 in place of each x and simplify, you find that the two sides of the equation are not equal:

$$5x + 3 = 4x + 10$$
$$5(9) + 3 \stackrel{?}{=} 4(9) + 10$$
$$45 + 3 \stackrel{?}{=} 36 + 10$$
$$48 \neq 46$$

When dealing with equations that have the same variable in more than one place, your child may wonder whether the variable has to be the same number each time. The answer is yes: You must substitute the same number of x on the left side of this equation and on the right. If the variables were meant to represent different values, then two different variables would be used, like x and y.

Your child will very likely encounter problems that ask whether a certain number is a solution to a given equation. This is a great time to ask your child what she understands. "How can you find out whether it is a solution or not?" or "What does it mean for it to be a solution?" If the response is an "I don't know," take the opportunity to dig in—before she gets any further into this topic. Does "I don't know" mean "I really don't understand what this stuff means," or is it "I don't get it, I give up"? Either way, this is not a time to give up or leave any uncertainty alone, because if there's anything she doesn't understand at this point, it's going to get a lot harder very quickly.

Solving One-Step Equations

Once your child understands what a solution means, the next thing to do is start solving some equations. Here are examples of equations that only have one operation, and therefore only take one step to solve.

1. $x + 1 = 5$

2. $x - 3 = 10$

3. $6x = 18$

4. $\frac{1}{2}x = 5$

In 6th grade, your child will be expected to master the skill of solving these types of equations, where x undergoes a single operation, and the result is given. In all likelihood, your child already solved lots of problems like these in elementary school, but instead of using x, there may have been a "?" or a box:

$$7 + ? = 9$$

$$4 \times \square = 32$$

Most students find solutions to problems like the ones above fairly easily and do not need a consistent system for showing their work. However, when the equations become more advanced in 7th grade and beyond, it will be important to have a foundation to build on. It is never too early to help your child practice good algebraic habits. Let's see how we might solve the equations above.

WARNING

If you follow the advice in this section and demand that your child show her work to these one-step problems, you are likely to hear "Leave me alone!" Most students are able to easily solve these equations in their head, and many do not see a point in showing their work, which takes extra time. Continue to press your child to show her work—it will make a big difference later.

Examples

To solve each of these equations we listed previously, we can use a number of different strategies.

1. $x + 1 = 5$

One way is to think of the equation in words. The left side says "one more than x." So 5 is one more than x. That should make it easy: $x = 4$.

Another way is to use fact families, which we discussed in Chapter 1. The fact family for this equation is:

$$x + 1 = 5$$
$$1 + x = 5$$
$$5 - x = 1$$
$$5 - 1 = x$$

The second and third equations are no easier to solve than the original, but the last one makes everything simple, because we know what $5 - 1$ is.

These first two methods are simple to perform mentally—your child is likely already doing them.

A third way, which will help your child for the coming years, is to reverse the operation in the equation in order to get x alone on one side. In the original equation, x is accompanied by $+1$. To get rid of that, we must do the opposite—subtract 1.

Since both sides of an equation have the same value, any operation we perform to change one side must also be performed on the other to keep everything equal. For example, we could subtract 2 from both sides in the equation $5 + 7 = 12$. We'd see that $5 + 7 - 2 = 12 - 2$, which is true. Similarly, let's subtract 1 from both sides of our algebraic equation:

$$\begin{aligned} x + 1 &= 5 \\ -1 \;\; &-1 \\ x &= 4 \end{aligned}$$

Some students find it helpful to draw a line straight down through the equals sign to keep the two sides of the equation separate. This will help later on, when there are variables on both sides of the equation.

NOTE

This method is one that your child's high school teachers will likely expect him to be familiar with, and it is the one that can be used for the widest variety of problems, from the simplest to the most complicated. Encourage your child to show his work in this way, even if he is more comfortable with other methods or uses other methods in his head.

This is a good time to ask your child some of those *why* questions we discussed in Chapter 7. See how well your child understands the method that he is using here. What is the point of subtracting 1? Why do we have to do it on both sides of the equation?

2. $x - 3 = 10$

$$\begin{aligned} x - 3 &= 10 \\ +3 \quad & +3 \\ x &= 13 \end{aligned}$$

3. $6x = 18$

The coefficient tells us that x is being multiplied by 6. To reverse that operation, we divide both sides by the coefficient, 6:

$$\begin{aligned} 6x &= 18 \\ \div 6 \quad & \div 6 \\ x &= 3 \end{aligned}$$

4. $\frac{1}{2}x = 5$

Multiplying x by $\frac{1}{2}$ is equivalent to dividing by 2. Therefore, one way to solve this equation is by multiplying both sides by 2.

$$\begin{aligned} \tfrac{1}{2}x &= 5 \\ 2 \cdot (\tfrac{1}{2}x) &= 2 \cdot (5) \\ x &= 10 \end{aligned}$$

Students may also solve this by *dividing* both sides by $\frac{1}{2}$, which is equivalent to what we have already done. $5 \div \frac{1}{2} = 5 \times 2 = 10$.

Solving Proportions

In Chapter 4, we worked with proportions and learned a variety of ways to solve them. At that time, we avoided the classic cross-multiplication method because it is so often taught as a procedure without making sense of why the math works. Now we will look at cross-multiplication and examine why it makes sense, in terms of what we have just seen for one-step equations.

Let's begin with a simple example, where one side of the equation is a whole number:

$$\frac{x}{4} = 9$$

RECOGNIZING DIFFERENT FORMS

As always, the fraction in this equation indicates division: x is being divided by 4. There are a few different ways this could be written, all of which are equivalent:

$$\frac{x}{4} = 9$$

$$x \div 4 = 9$$

$$\frac{1}{4}x = 9$$

This last form confuses many students at first because we have taken the x out of the fraction. Remind your child that dividing a number by 4 is equivalent to multiplying it by $\frac{1}{4}$.

All we need to do to isolate x is multiply both sides by 4:

$$4 \bullet (\tfrac{x}{4}) = 4 \bullet (9)$$

$$x = 36$$

When an equation has a fraction on both sides (which is what we usually mean when we talk about **proportions**) and the variable is in a numerator, it is just as simple:

$$\frac{x}{3} = \frac{6}{8}$$

$$3 \bullet (\tfrac{x}{3}) = 3 \bullet \tfrac{6}{8}$$

$$x = \frac{18}{8}$$

We can reduce this last fraction to $\frac{9}{4}$, express it in mixed-number form as $2\frac{1}{4}$, or convert it to decimal form, 2.25.

If the variable has a coefficient that is a non-unit fraction, it takes a bit more arithmetic to solve:

$$\frac{2}{3}x = 12$$

Here we are multiplying x by $\frac{2}{3}$, which is another way of saying "multiplying by 2 and dividing by 3." We begin by eliminating the denominator:

$$3 \cdot (\tfrac{2}{3}x) = 3 \cdot (12)$$

$$2x = 36$$

$$\div 2 \quad \div 2$$

$$x = 18$$

Where things get tricky is when the variable is in a denominator.

$$\frac{6}{x} = \frac{5}{2}$$

MISCONCEPTION ALERT!

In order to solve this, we need to get the x out of the denominator. Many students do not realize how important this step is. They think that if they can get rid of the 6 (which they do by either multiplying or dividing by 6, since they are not clear on the meaning of the equation), then the x will be alone and the equation will be solved. This is not true. If we were to simply make the 6 vanish, the x would still be in the denominator of a fraction and we would be left with $\frac{1}{x}$, which is VERY different from x.

Let's get x out of the denominator first. It doesn't matter which denominator we eliminate first—ultimately we want to get rid of both fractions.

$$x(\tfrac{6}{x}) = x(\tfrac{5}{2})$$

$$6 = \tfrac{5}{2}x$$

Now we multiply both sides by 2:

$$2 \cdot (6) = 2 \cdot (\tfrac{5}{2}x)$$

$$12 = 5x$$

And divide by the coefficient:

$$\tfrac{12}{5} = \tfrac{5x}{5}$$

$$x = \tfrac{12}{5} = 2.4$$

To summarize, what we have done is multiply both sides by both denominators. Each side's original denominator canceled out, so what we were left with on each side was the product of one numerator and the opposite denominator. In other words, the 6 and the 2 multiplied on one side, and the 5 and the *x* multiplied on the other side. Do you see why this is confusing? This is *exactly* what happens when we cross-multiply:

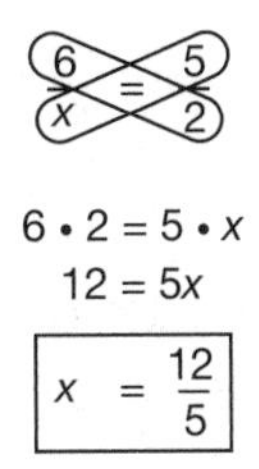

Hopefully, taking it step by step will help you and your child solve proportions in a more sensible way than cross-multiplication.

Solving Two-Step Equations

And now we come to those classic "solve for *x*" problems.

So far, we have been looking at one-step equations, where *x* was either multiplied, divided, added, or subtracted to or from one other number to give a known result. Now we will be solving equations in a variety of forms, starting with two-step equations where multiplication/division and addition/subtraction are both involved.

There is a lot of potential for *I don't knows* in this topic. Many students approach this topic expecting it to be easy, so they might see an equation like $2x + 1 = 7$ and instantly figure out that *x* is 3 without much thinking—without, in fact, even being able to show or explain how they got the answer. It's just that simple for them, and that's not a bad thing. The danger is when it comes to an equation like

$$24x - 67 = 29$$

Although this problem also has a simple solution, most students (and their parents) are unlikely to figure it out mentally. Here, many students will throw an *I don't know* out there at this point, hoping that it will excuse them from having to work for an answer. Press your child if she does this, and help her see how these two examples are more similar than different—and they are both equally simple.

Reversing the Order of Operations

Since we are already experts at writing and interpreting algebraic expressions, let's use this skill to understand what an equation is telling us. $2x + 1 = 7$ means that the number x represents is multiplied by 2 and 1 is added to the product, with a final result of 7. The order of operations is very important because it tells us that the number is multiplied by 2 *before* 1 is added, not after.

In one-step problems, we reversed the operation to solve for x. Now that we have two steps, we must reverse each operation, but we must also reverse the order of operations. That is, if we multiplied by 2, then added 1, then the reverse would be to subtract 1, then divide by 2:

$$2x + 1 = 7$$

$$-1 \quad -1$$

$$2x = 6$$

$$\div 2 \quad \div 2$$

$$x = 3$$

In words, we could reason that if $2x + 1 = 7$, then $2x$ must be 6, and if $2x = 6$, then x must be 3.

The same method works for $24x - 67 = 29$. x is first multiplied by 24, and then 67 is subtracted from the result, leaving 29. To reverse it, we start with 29, add 67, and then divide by 24:

$$24x - 67 = 29$$
$$+67 \quad +67$$
$$24x = 96$$
$$\div 24 \quad \div 24$$
$$x = 4$$

CHECKING YOUR SOLUTIONS

Whenever you solve an equation, it is a good idea to check that your solution is correct. You can do this by substituting the solution (the value you have found for the variable) into the original equation. If you work out the order of operations and both sides of the equation come out equal, then you have the correct solution. For example:

$$24x - 67 = 29$$
$$24(4) - 67 \stackrel{?}{=} 29$$
$$96 - 67 \stackrel{?}{=} 29$$
$$29 = 29$$

Since both sides of the equation are the same, our solution of 4 is correct.

When solving equations by this method, it is very important to show that you are performing the same operations on both sides of the = sign. An equation tells us that the expressions on both sides of the = sign have the same value. If you make a change to one of those expressions, you must make the same change to the other—or else they will no longer be equal.

To illustrate this idea, suppose x is 3. There are lots of different equations we could use to say that:

$$x = 3$$
$$x + 1 = 4$$
$$4x = 12$$
$$4x + 5 = 17$$

All of these equations are equivalent. In each case, the same change has been made to both sides of the equation. You can add the same number to both sides, or multiply both sides by the same number without changing the overall meaning of the equation.

In the previous example, the equations $2x + 1 = 7$, $2x = 6$, and $x = 3$ are all equivalent. By performing the inverse of each operation, in the reverse order of operations, we stripped the original equation down to its simplest form, which tells us exactly what x is.

Examples

1. $33 = 12x - 27$

The first time your child sees a problem like this, you may hear another "I don't know." Many students have trouble with equations like this at first because the variable is on the other side. Students may be used to seeing operations on the left and an "answer" on the right. It is important to help your child see that an equation is a statement that two things have the same value, whether they are numbers, expressions, or a combination of the two. Operations may happen on either side of the equation.

$33 = 12x - 27$	Original equation
$+27 \quad +27$	Add 27 to both sides
$60 = 12x$	New equation
$\div 12 \quad \div 12$	Divide both sides by 12 (the coefficient)
$5 = x$	Final solution

2. $\frac{b}{5} + 14 = 9$

In this example, b is being divided by 5, instead of multiplied by a coefficient. Otherwise, it is no different—it just means our final step will be multiplication, rather than division.

$\frac{b}{5} + 14 = 9$	Original equation
$-14 \quad -14$	Subtract 14 from both sides
$\frac{b}{5} = -5$	New equation
$\cdot 5 \quad \cdot 5$	Multiply both sides by 5 (the divisor)
$b = -25$	Final solution

3. $58 - 3x = 25$

This example looks different from the previous ones because the x term is not the leading term on its side, and it is being subtracted. As a teacher, this is a time when I *hope* to hear an "I don't know" from my students because it shows that they recognize the difference. The alternative is students assuming that they do know, which in this case is usually going to lead to a misconception. Students will often add 58 to both sides, reasoning that addition is the reverse of subtraction. But it is not the 58 that is being subtracted, so that is not the correct way to deal with this equation. Subtracting 58 from both sides should leave you with $-3x$ on the left side. But when students are thinking in this way, they usually write $3x$, so that their solution ends up having the opposite sign of the correct solution.

There are two other methods that work well.

Method #1: Turn Subtraction to Addition

$58 - 3x = 25$	Original equation
$58 + (-3x) = 25$	Change subtraction to addition
$-3x + 58 = 25$	Use the Commutative Property to put the x term in front. Now it looks more like what we're used to.
$-58 \quad -58$	Subtract the constant term from both sides
$-3x = -33$	This is equivalent to the original equation
$\div -3 \quad \div -3$	Divide both sides by the coefficient of x
$x = 11$	We have our solution.

Method #2: Move the x Term

$58 - 3x = 25$	Original equation
$+3x \quad +3x$	Add $3x$ to both sides. This will make the coefficient of x positive on the right-hand side. Be careful not to combine unlike terms.
$58 = 3x + 25$	New equation
$-25 \quad -25$	Subtract the constant term from both sides
$33 = 3x$	New equation
$\div 3 \quad \div 3$	Divide both sides by the coefficient
$x = 11$	We have our solution—the same one we got from the other method.

Simplifying Equations

So far, all of the equations we have solved have been in the form $mx + b = c$. The letters m, b, and c all represent constants—and the equations may look a little bit different, but they all boil down to the same thing, where x is multiplied by something (m), and then something else (b) is added to give a specific result (c). The values of m, b, and c could be anything—positive, negative, zero, a whole number, or a fraction.

Sometimes your child will encounter more complicated equations, where more than two operations are happening, or where parentheses change the order of operations, or where the variable may appear in more than one place. In these cases, our first goal should be to convert the equation to $mx + b = c$ form, at which point it is a simple two-step solution.

More than Two Terms on One Side

Consider the equation $5x + 7x - 3 = 45$. In this equation, there are two terms with the variable x. We could eliminate the constant term –3, but before we can completely solve this equation, we need to find the total number of x's on the left side.

$5x + 7x - 3 = 45$	Original equation
$12x - 3 = 45$	Combine like terms ($5x + 7x = 12x$)
$+3 \quad +3$	Eliminate the constant term
$12x = 48$	New equation
$\div 12 \quad \div 12$	Divide to eliminate the coefficient
$x = 4$	Our solution

Encourage your child to always check his work after solving an equation. So, in this case, we will plug 4 in for x in the original equation. Make sure that your child always goes back to the *original* equation—the one printed in the book or on the worksheet. If you substitute into any other equation that you've worked out along the way, then your check will not catch any errors you made up to that point.

For this example, we check by substituting 4 for x in $5x + 7x - 3 = 45$:

$$5(4) + 7(4) - 3 \stackrel{?}{=} 45$$
$$20 + 28 - 3 \stackrel{?}{=} 45$$
$$48 - 3 \stackrel{?}{=} 45$$
$$45 = 45$$

Your child might prefer to do this check mentally, if it is faster. That's fine, as long as she is checking.

Parentheses

Sometimes an equation will contain parentheses, such as $5(3x - 1) = 115$. There are lots of ways to solve any equation, but again, one method that your child should definitely master is by converting it to $mx + b = c$ form. We can eliminate the parentheses by distributing the 5.

$5(3x - 1) = 115$	Original equation
$15x - 5 = 115$	Distribute the 5. Now we have $mx + b = c$ form.
$+5 \quad +5$	Eliminate the constant term
$15x = 120$	New equation
$\div 15 \quad \div 15$	Divide by the coefficient
$x = 8$	Our solution

Take a moment to check this solution for yourself.

Sometimes an equation requires both the Distributive Property and combining like terms. For example, $\frac{2}{3}(9 - 21x) + 11x = -24$. Since the order of operations tells us to multiply before we add, the Distributive Property should be performed before combining like terms.

$\frac{2}{3}(9 - 21x) + 11x = -24$	Original equation
$6 - 14x + 11x = -24$	Distribute the $\frac{2}{3}$
$6 - 3x = -24$	Combine like terms. Since the $14x$ is negative, we end up with $-3x$.
$6 + (-3x) = -24$	Change the subtraction to addition
$-3x + 6 = -24$	Apply the Commutative Property. We now have $mx + b = c$ form.
$-6 \quad -6$	Eliminate the constant term
$-3x = -30$	New equation
$\div -3 \quad \div -3$	Divide by the coefficient
$x = 10$	Our solution

Take a moment to check this solution.

Variables on Both Sides

Sometimes you will see equations where the variable appears on both sides of the equals sign, like $3x + 9 = 5x + 1$. Students are often confused by this type of equation because they are accustomed to seeing an "answer" on the right side of the equals sign. They see the $5x + 1$ as a separate problem to be solved, unrelated to the $3x + 9$. Help your child think of the equals sign as the center of a balance scale.

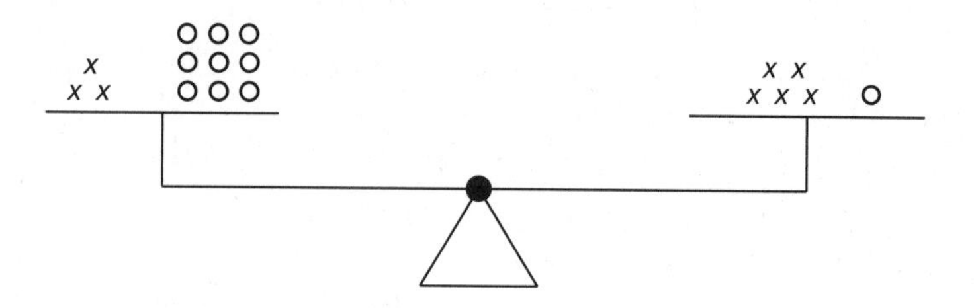

In this picture, we can see how both sides are balanced. We can add or remove quantities from either side of the scale, as long as we treat both sides the same way to maintain the balance. So we can remove the 3 x's from the left side if we also remove 3 x's from the right side as well.

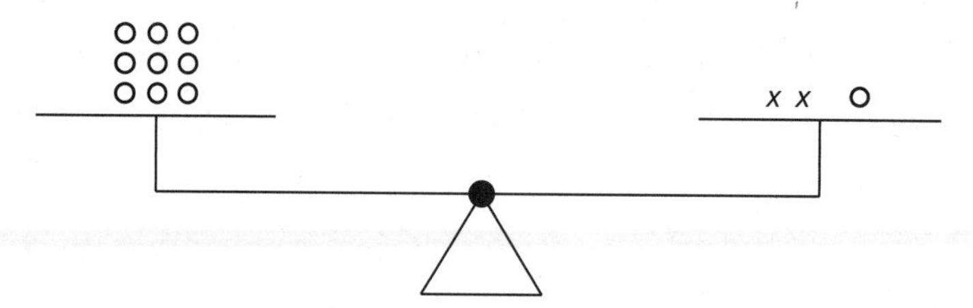

Now we can rewrite the equation in an equivalent form:

$$9 = 2x + 1$$

This is now a straightforward $mx + b = c$ equation.

We can show the same process in equation form:

$3x + 9 = 5x + 1$	Original equation
$-3x \qquad -3x$	Subtract $3x$ from both sides
$9 = 2x + 1$	This is now $mx + b = c$ form.

You could also subtract the $5x$ in the first place, which would result in a negative x term on the left: $-2x + 9 = 1$. It is still equivalent to the other forms and will result in the same solution, but it is preferable to deal with positive numbers. Eliminating the smaller coefficient will always leave you with a positive coefficient.

The balance model becomes a little tricky when dealing with subtraction. How do you represent negative weight? For this reason, it is best to guide your child toward the method of manipulating equations.

For example:

$20 - 6x = 2x - 52$	Original equation
$+ 6x \quad + 6x$	Add $6x$ to both sides to eliminate the $-6x$
$20 = 8x - 52$	Now we have $mx + b = c$ form.
$+ 52 \qquad + 52$	Eliminate the constant term
$72 = 8x$	New equation
$\div 8 \quad \div 8$	Divide by the coefficient
$9 = x$	Our solution

Summary

In this chapter, we have tackled one of the most important and common topics in algebra: solving equations. There is no doubt that your child will require a great deal of practice on the skills we have used here. Very few students master it instantly. The conceptual reasons for each step are often lost in the process, and the more your child practices, the better he will be able to interpret what is happening in each step and what the algebraic symbols represent.

As your child is practicing, ask him to explain the reason for each step he is doing. If your child is not sure how to explain, or simply says, "I don't know," keep pushing him toward the idea of creating simpler equivalent equations in order to eventually get the variable by itself: $x = \_\_$.

As a parent trying to help your child, you may get very frustrated hearing "I don't know." Know your child, be patient, and always remember that you love each other. Don't let math get in the way of that. Every *I don't know* is a door—it can be either open or closed. Sometimes your child may try to close the door to learning with an *I don't know*. As the parent, practice interpreting whether that door is shut permanently or just for the moment. If it's just for now, do something to change the situation, break the mood, and come back to it later with a clear mind. If it's more permanent, then you need a bigger solution. Consider meeting with your child and her math teacher to come up with a plan. Be persistent, and let your child know all the time that you know this work is hard, you believe in her, and you know she can learn it. It's going to take a lot of hard work, but you are willing to help. You can be both coach and cheerleader.

In Chapter 9, we will move on to our final algebra topic: linear equations. This is as abstract as it gets for middle school math, yet in many ways it is the most relevant to real life. Throughout the chapter, we will make many connections to day-to-day life, which will make the math both easier to understand and more interesting.

Math Is Life

Linear Relationships in Everyday Life and Why They Matter

Linear Relationships in Life

In Chapter 8, we solved equations like $5x - 9 = 41$. Equations of this kind are known as **linear equations** because they describe a pattern that can be shown as a line on a graph. You may remember the general form $y = mx + b$ from your school days. In this chapter, we will work with $y = mx + b$ equations and see the many ways they connect to our everyday lives.

The eternal question, "When are we ever gonna need this?" comes up more often for some topics than it does for others. This topic is one that makes that question easy for a teacher to answer because linear relationships pop up so often in life.

What Is a Relationship?

When mathematicians use the word **relationship**, they mean something different from what we usually think. Remember that a variable can be used to represent a quantity that changes, such as the temperature outside, the amount of money in your bank account, or your child's height. We say that two variables have a relationship if a change in one of the variables is associated with a change in the other variable. For example, the more electricity you use, the more your utility bill will cost. If your child does more homework, then his grade will likely increase. As you spend more time exercising, you might lose weight, or gain strength, or become healthier.

Sometimes a relationship can be inexact. It is impossible to predict exactly how much weight you will lose for every minute you spend on a treadmill. Other times there is a pattern, but it may take some sophisticated math to calculate—like trying to determine what time the sun will rise and set on a certain day based on the time of year.

Then there are relationships where a simple formula tells you everything you need to know. When you buy movie tickets for your family, each ticket costs a certain amount, so the total is easy to find. If you have a limited cell phone plan, then your monthly bill likely includes a certain monthly fee and a certain charge per minute or per text. These are examples of **linear relationships**, which we will spend much of this chapter studying.

Proportions: The Simplest Linear Relationships

In Chapter 4, we defined a proportion as *an equation that shows two equivalent ratios*. Now we will extend that idea to understand a proportion as a relationship between two variables that change. For example, when you take your car to the gas station, the two counters showing gallons and dollars are proportional to one another. Every gallon costs a certain number of dollars, and for every dollar, you get the same amount of gas. We have seen and used proportions in this book when dealing with maps, scale drawings, and unit conversions, such as converting kilograms to pounds.

One skill your child will be expected to master is recognizing when a relationship is proportional. We'll look at one example of a relationship that is proportional and another that is not.

Examples

1. One example where you may be looking for a proportional relationship is when comparing prices of two different sizes of the same product at the grocery store. Suppose a 12-ounce box of your child's favorite cereal costs $3.99. Let's call it $4. If a 20-ounce box costs $4.99, or say $5, is this relationship proportional?

   Let's look at the ratio of weight to price for each package. The 12-ounce box is $\frac{12 \text{ ounces}}{\$4}$ and the larger box is $\frac{20 \text{ ounces}}{\$5}$. Now we must check to see whether these fractions are equivalent—in other words, does $\frac{12}{4} = \frac{20}{5}$? We saw a number of ways to answer this question in Chapter 4, but in this case both fractions can easily be reduced to a unit rate: $\frac{12}{4} = 3$, and $\frac{20}{5} = 4$. So the smaller box gives you 3 ounces per dollar, while the larger box contains 4 ounces per dollar. Therefore, these values are *not* proportional. We say that the cost is *not* proportional to the size of the box. If you have the cash to buy the larger box, it will end up saving you money in the long run. This is a very simple example of how proportional reasoning can help you save money.

2. Kayla and Angelina are doing a science project. They measure each other's height and the length of their shadows at a certain time of day. The data is recorded below. Are the girls' shadows proportional to their heights?

	Height (cm)	Shadow Length (cm)
Kayla	183	244
Angelina	162	216

   We can set up an equation comparing the ratios of each girl's height to her shadow length: $\frac{183}{244} = \frac{162}{216}$. Both fractions can be reduced to $\frac{3}{4}$. Therefore, shadow length is proportional to height. We could use this ratio of $\frac{3}{4}$ to make predictions about how long another person's shadow would be at the same time of day, based on her height. While measuring shadows may not be the kind of thing you and your child do every day, many middle school science teachers assign this activity.

Coordinate Graphs

Graphs are one of the most useful and commonly used ways of representing a relationship. They are often the best way to compare two different things. Graphs are used and misused all the time in the media, so being able to read and interpret a graph is not only an essential math skill but also an essential life skill.

A graph has two **axes**, or number lines that cross. The ***x*-axis** is horizontal, with smaller numbers to the left, and larger numbers to the right. The ***y*-axis** is vertical, with larger numbers above, and smaller numbers below. Points on the graph are identified by their **coordinates**—their locations relative to both the *x*- and *y*-axes. For example, point *A* in the graph below is (4,–1). Its *x*-coordinate is 4, and its *y*-coordinate is –1. Point *B* is (–3,0).

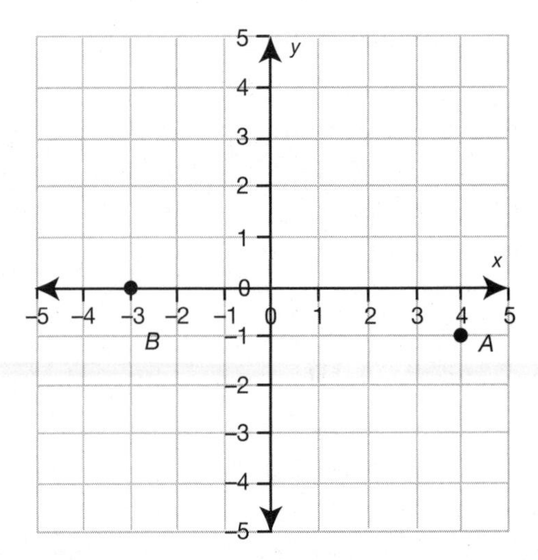

A coordinate graph is used to show a relationship between two variables, like *x* and *y*. In the following graph, *y* represents how much money Valecia has saved from her after-school job, and *x* represents how long she has been saving.

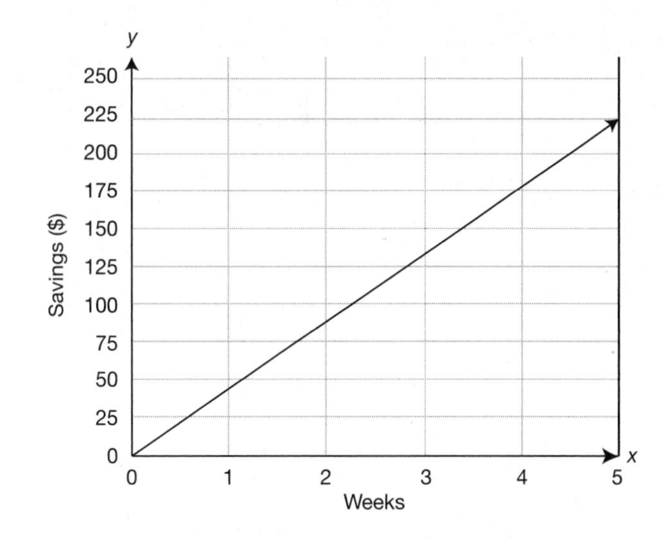

If Valecia wants to know how much money she will have saved after a certain number of weeks, she can find that information on the graph. For instance, after 4 weeks, you can see that she has saved a little more than \$175. If you find 4 on the x-axis and look straight above it until you hit the graph line, you can see that the savings is just above the \$175 line.

Graphing Proportions

Graphs are a powerful tool for making comparisons and predictions when you have incomplete information about a situation. In the shadow length example from before, we could use height and shadow length as our two variables. If we label our x-axis as Height, and our y-axis as Shadow Length, then Kayla's coordinates are (183,244), and Angelina's are (162,216). These points on the graph can be used to represent the two girls.

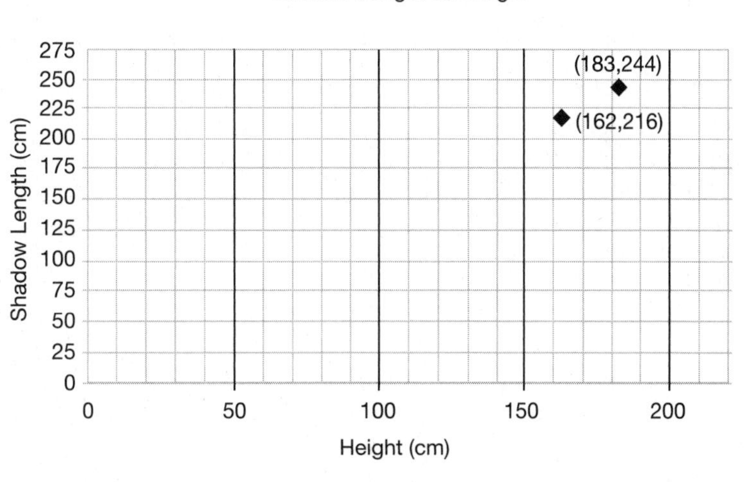

Now we can draw a line through the two points:

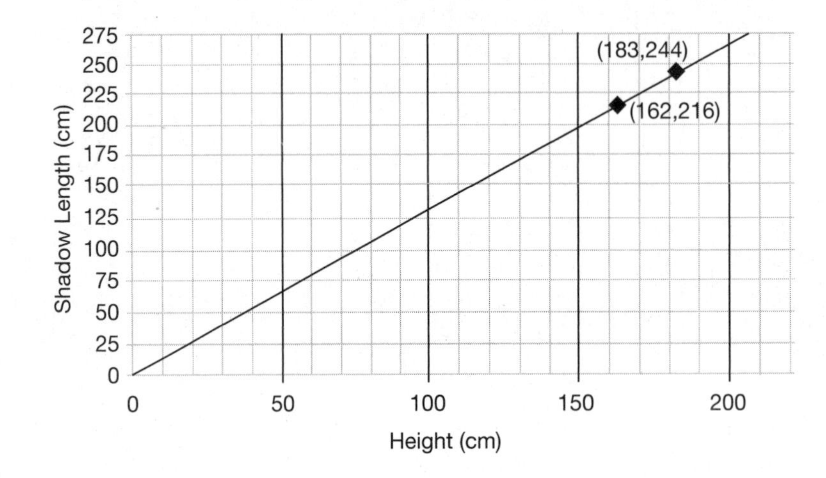

This line represents the proportional relationship between shadow length and height. Because these two variables have a definite mathematical relationship, any point on the line tells you the shadow length of a person of a certain height. For instance, since the line passes through (150,200), we know that a person who is 150 cm tall will have a shadow that's 200 cm long.

We can use this line to make predictions about people of different heights. If you want to know how long a person's shadow would be, simply find her height on the *x*-axis, move your finger straight up to the line, and find the *y*-coordinate of that point. The graph below shows this process for a person 110 centimeters tall.

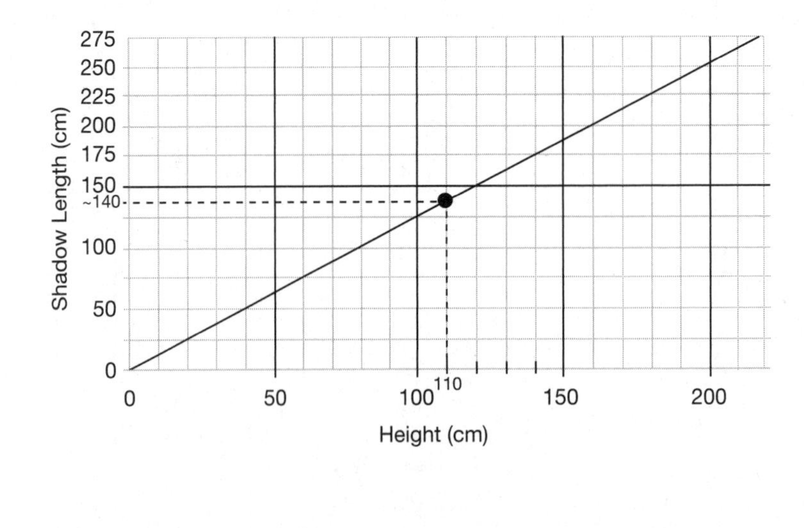

A person 110 centimeters tall will have a shadow about 140 centimeters long.

Notice that the line in the shadow length graph passes through the origin, (0,0). This is a feature of proportional relationships: They include the point (0,0). If an object were 0 centimeters tall, it will have a shadow 0 centimeters long. The same is true for any proportional relationship. If you buy 0 gallons of gas, you pay $0.00. If you bake 0 cookies, you will need 0 cups of sugar.

The Constant of Proportionality in a Graph

In Chapter 4, we discussed the constant of proportionality, which—like a unit rate—tells us what to multiply one variable by in order to find the other value in a given relationship. For example, if you are buying deli meat that costs $4.49 per pound, you can multiply the number of pounds by $4.49 to find the total cost.

A graph can be used to find the constant of proportionality for any proportional relationship. The graph below is something you might see while exercising at the gym.

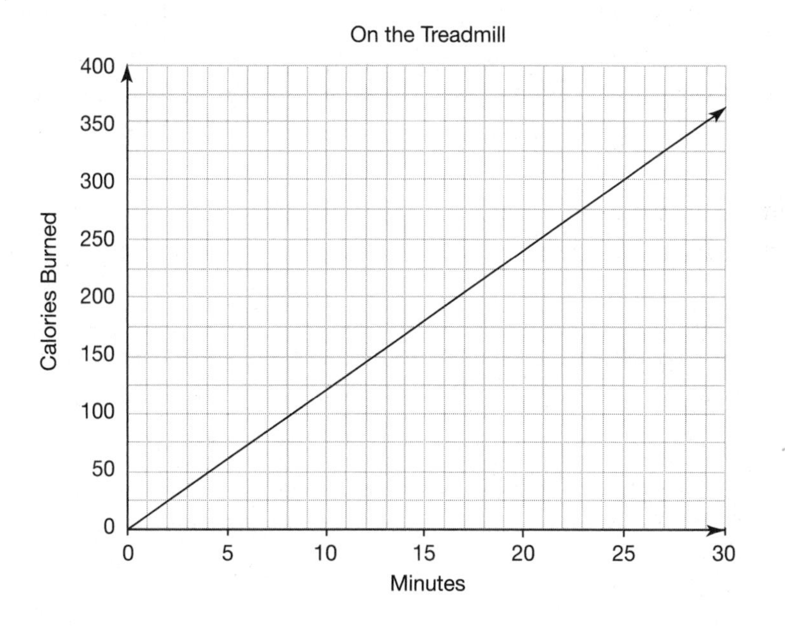

First, note that the line in this graph is straight, and that it passes through (0,0)—that is, after 0 minutes, we have burned 0 calories. Since the line passes through (0,0), the relationship is proportional. The graph shows that the more time we spent on the treadmill, the more calories we've burned—and each minute we burn the same number of calories. (In real life, we are never this consistent—we burn more calories in some minutes than others; when we say "math is life," we really mean "math is a model that can be used to represent life, but it is rarely a perfect representation.")

We can use any point on this line to find the constant of proportionality. A convenient point to use is (25,300) because the line passes perfectly though a corner of the grid. After 25 minutes on the treadmill, the runner has burned 300 calories. If we divide, that gives us our unit rate: $\frac{300}{25}$ = 12 calories per minute. As we have seen, we can use this constant of proportionality for any other value—on or off—the graph. If a runner stays on the treadmill for 45 minutes, for example, she will burn 45 × 12 = 540 calories.

MISCONCEPTION ALERT!

Dividing the *y*-coordinate of a point by the *x*-coordinate to find the constant of proportionality works only if you know for certain that the graph *is* a proportion. If the line does not pass through the origin (0,0), then it does not represent a proportion, and this method does not work because *there is no constant of proportionality*. Later in this chapter, we will discuss the concept of slope, which replaces the constant of proportionality in graphs that do not represent a proportional relationship.

Comparing Graphs and Constants of Proportionality

The following graph shows the same treadmill data, with a second line added in. The solid line is our old data, with the person burning 12 calories per minute. The dotted line below it represents someone who is burning 10 calories per minute.

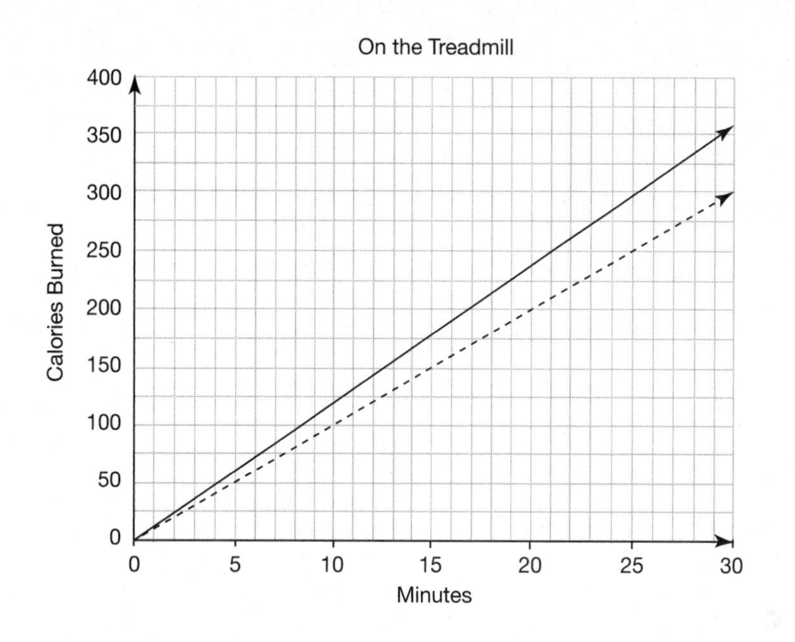

As you can see, at 0 minutes, both lines start off at 0 calories. As time goes on, however, the gap between the lines grows larger and larger. The person who burns 12 calories per minute constantly gains on the person who burns 10 calories per minute, getting farther and farther ahead on their calorie count with each minute that passes.

Examples

1. The graph below shows how far a car has traveled after a certain number of hours on the road. As we have seen before, this is not a completely accurate representation of the real world because the driver would have to stop periodically and traffic and other conditions would not allow a real car to travel for so long at a constant speed. That being said, let's interpret what we can from the graph.

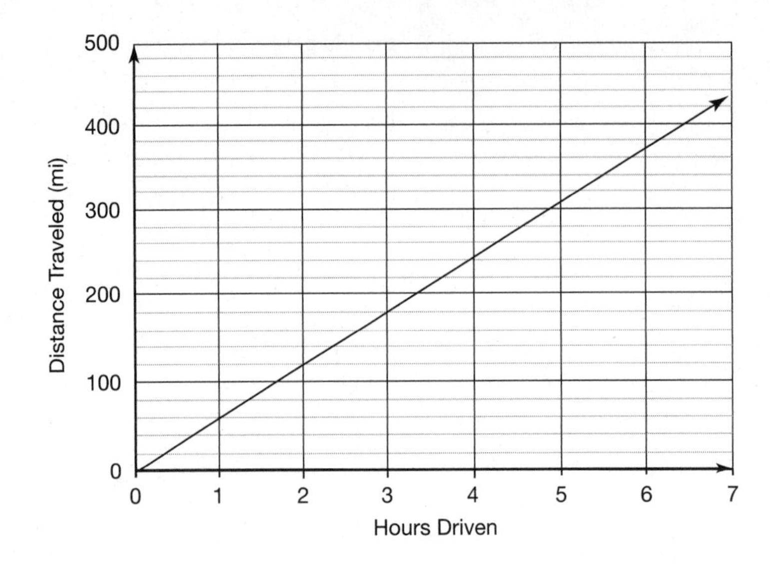

The first thing to notice is that this is a proportional relationship, since it passes through (0,0). After 0 hours, the car has traveled 0 miles.

Next, let's figure out the unit rate, or constant of proportionality, for this graph. The points (0,0), (1,60), (2,120), etc. are on the graph. These points follow a very specific pattern: For every hour that passes, the distance traveled increases by 60 miles. Therefore, we have a unit rate of 60: The car is moving at 60 miles per hour.

There is another way to find the constant of proportionality in this graph, which will be very important later on. In the diagram below, several right triangles (which we will refer to later as **slope triangles**) are drawn to show the unit rate in some of its different forms.

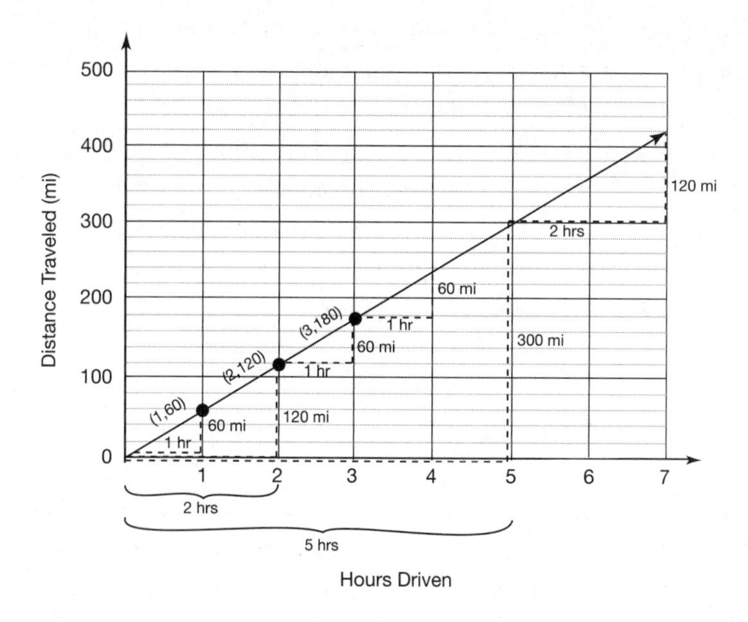

Remember similarity from Chapter 5? All of these triangles are similar to one another. Their height and base length all have the same ratio:

$$\frac{60 \text{ mi}}{1 \text{ hr}} = \frac{120 \text{ mi}}{2 \text{ hr}} = \frac{300 \text{ mi}}{5 \text{ hr}} = 60 \text{ mph}$$

2. Here is another example about a favorite topic among middle schoolers: money.

In this graph, it is a little bit more difficult to find a unit rate because at 1 hour the graph line passes *between* exact grid points. We can eyeball it and guess that it's about \$15 at 1 hour, but it could be \$14 or \$16 or \$13.75—it's better to look for places where the line crosses *perfectly* through a corner of the grid. This happens at 2 hours, when the amount earned is \$30.

$$\frac{\$30}{2 \text{ hours}} = \$15 \text{ per hour}$$

We can use other points to confirm—such as 4 hours and \$60 or 6 hours and \$90, which both give the same unit rate.

3. If you and your child like to bake cookies, you might already have an idea of how much butter goes into a dozen cookies. In this graph, it won't be so easy to estimate visually.

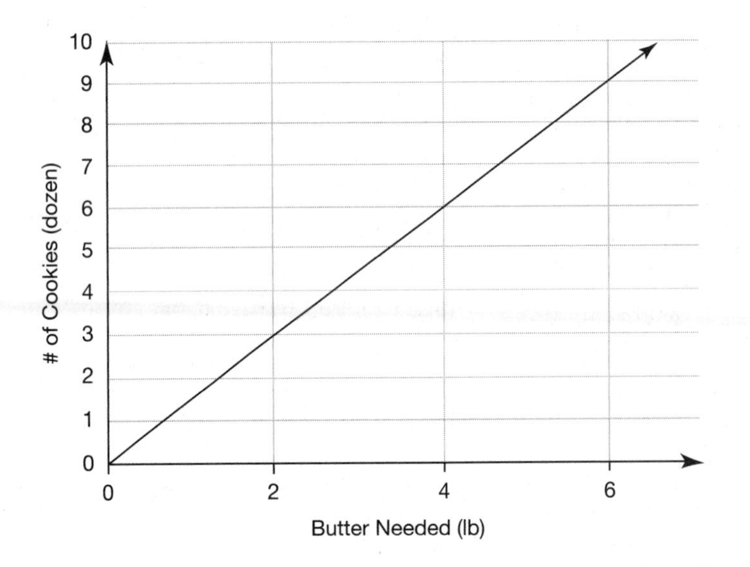

The first place after (0,0) where the line hits an exact grid point is (2,3)—that is, 2 pounds of butter for 3 dozen cookies. We can use this information to find a unit rate.

$$\frac{2 \text{ pounds butter}}{3 \text{ dozen cookies}} = \frac{2}{3} \text{ pound of butter per dozen cookies}$$

Multiple Representations

So far in this book, we have worked with four different representations for proportional relationships. Your child will see these same four representations over and over in all math courses from now on. She will be expected to learn how to convert between these forms efficiently and correctly.

- **Verbal description.** We have seen many verbal descriptions, such as "Tickets to the concert cost $10 per person." Verbal descriptions of a relationship are very important because they allow your child to make sense of what is happening with the numbers by relating them to a well-understood context.
- **Tables.** In Chapter 4, we worked extensively with ratio tables. A table can be a great way to see the relationship between two variables, especially when the numbers change in a predictable way—such as increasing by the same increment each time. The table below shows some of the numbers matching the verbal description "Tickets cost $10 per person."

People	0	2	4	6	8	10
Total cost ($)	0	20	40	60	80	100

  In this table, you can see the constant of proportionality ($10) if you divide any number in the bottom row (the total cost) by the associated number of people. Also notice that each time the number of people increases by 2, the total cost increases by $20, which is ten times as much. You can see in the first column of values that 0 tickets will cost $0.

- **Equations.** We also saw proportional equations in Chapter 4. The equation $y = kx$ represents a proportional relationship between the variables x and y, where k is the constant of proportionality. To continue with the $10 concert tickets, the equation would be $y = 10x$, where y stands for the total cost and x stands for the number of tickets. Of course, in the equation for a proportion, if $x = 0$ then $y = 0$ as well, since multiplying k by 0 will always result in 0.

- **Graphs.** This chapter has now introduced graphs, which take the two variables from the equation (or table, or verbal description), and turn them into a series of (x,y) coordinate pairs. In the graph below, each pair of values from the table has been turned into a set of coordinates. If you choose any point on this graph, the y value at that point (the total cost) will be 10 times as much as the x value (the number of tickets).

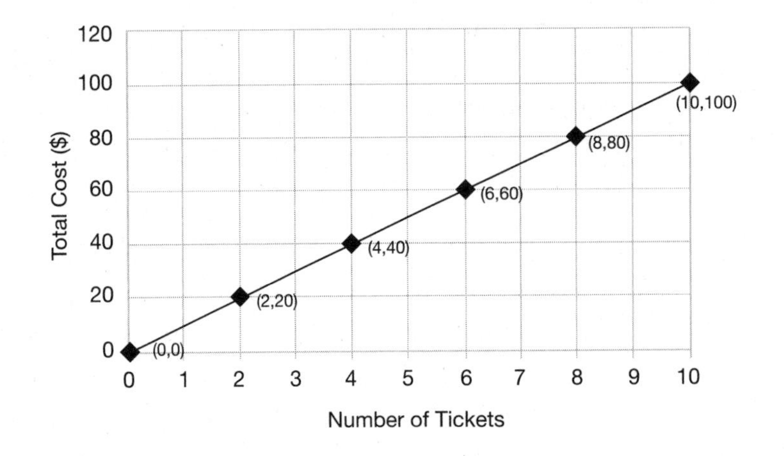

As your child is beginning to learn about these four representations, check to see how well she understands the connections. Does she see the numbers from a table in the graph that goes along with it? Can she explain how the constant of proportionality shows up in a table, graph, or equation? Does she understand how (0,0) is represented in each of the different forms?

Now we will move on to linear relationships, which are more general than proportions. Linear relationships have a lot of the same features as proportions, but with one key difference: They do not always pass through (0,0).

Linear Relationships

The graphs we have seen so far in this chapter all represent proportional relationships, which are extremely common in our day-to-day lives. But many graphs do not pass through (0,0). In real life, this happens because the starting value in a situation is not 0. For instance, when joining a gym, you might have to pay a one-time fee before your monthly rate kicks in. Some online stores offer a flat rate for shipping, which is added to the cost of your purchases. Your cable plan may have included an installation fee that you paid before your monthly charges started.

As our first example, suppose Elizabeth has decided to save up her money for a new necklace. When she started saving, she already had $45. Each week, she saves $5 from her allowance. Here is a graph for this situation:

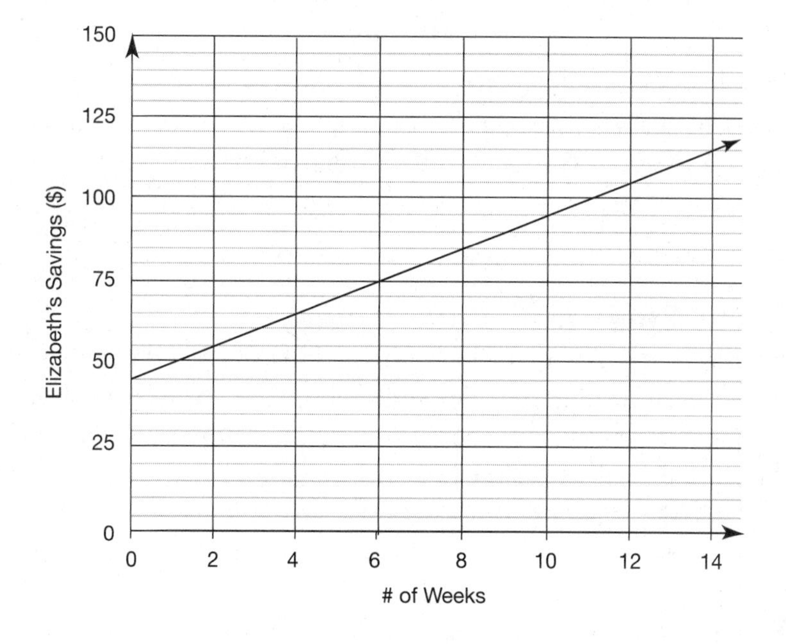

The big difference between this graph and the others we have seen is that the line does not pass through (0,0). Instead, it starts at (0,45) because when 0 weeks have passed, Elizabeth already has $45. We call this a **starting value**. This graph represents a **linear relationship** because the graph is a straight line. Any relationship that has a constant rate of change will form a straight line on a graph. In this case, the constant rate of change is Elizabeth's savings of $5 per week.

WHY A CONSTANT RATE OF CHANGE IS *NOT* A CONSTANT OF PROPORTIONALITY

When we worked with proportions, we found that for any point on the line, the ratio $\frac{y}{x}$ matched the constant of proportionality. In a linear graph, this will not be the case. If we divide the coordinates of any points on this graph, we will find that they have different ratios. For instance, Elizabeth had $75 after 6 weeks, and $115 after 14 weeks.

$$\frac{75}{6} = 12.5$$

$$\frac{115}{14} \approx 8.2$$

Since these ratios are different, the relationship is *not* a proportion. Even though Elizabeth is saving at a constant rate of $5 per week, you cannot simply multiply the number of weeks by $5 to find her total savings.

Rate of Change in a Linear Graph

How can we see the $5 that Elizabeth saves per week in the graph?

Earlier, we drew right triangles under the line of a proportional relationship to show the constant of proportionality. We can do a similar thing to see the constant rate of change in a linear graph.

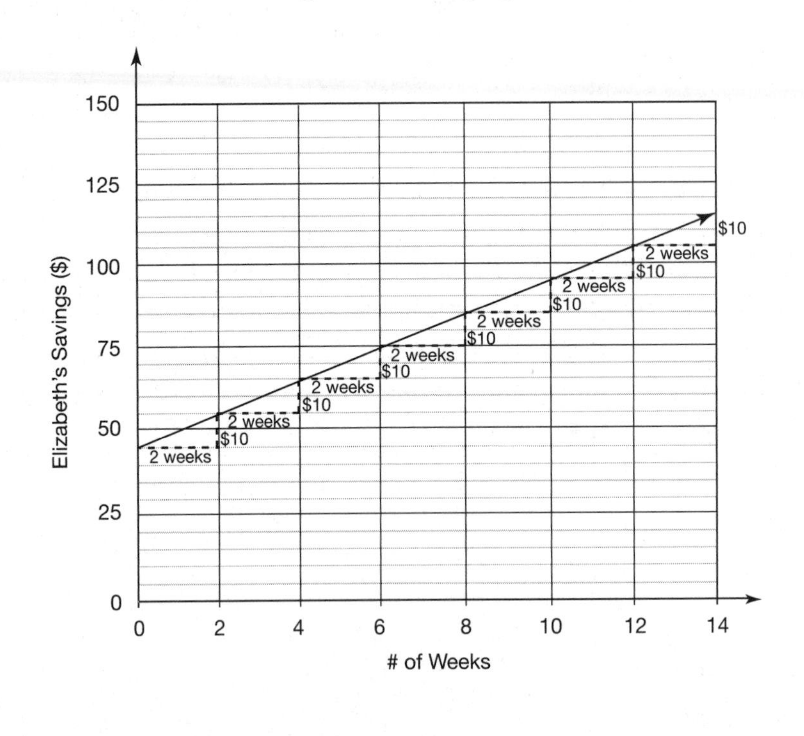

This graph shows a series of congruent right triangles. The horizontal leg of each triangle represents 2 weeks passing, and the vertical leg shows $10 earned. So the unit rate is $\frac{\$10}{2 \text{ weeks}}$ = $5 per week.

You can also use larger or smaller triangles—they will always simplify to the same ratio of $5 per week.

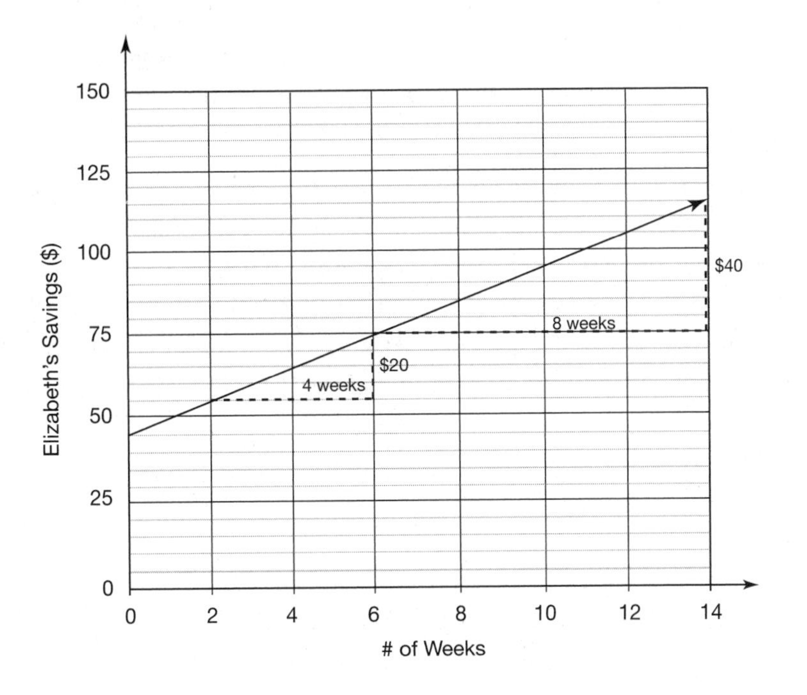

In the graph of a linear relationship, we call the rate of change the **slope**. The slope of the above graphs was $5 per week, meaning that for every unit (week) we move to the right on the x-axis, the y-coordinate on the line increases by $5.

What Is Slope?

Slope is a measure of how steep a line is. A line with a higher slope will be very steep, rising fast as you move to the right. A line with a small slope (like a fraction or decimal between 0 and 1) will have a slow rise. Slope tells you how quickly one variable changes along with the other variable in the relationship.

PROPORTIONS AND LINEAR RELATIONSHIPS

Any relationship whose graph is a straight line is considered a linear relationship. Since proportions have straight-line graphs, they are considered linear relationships. Just as every square is a rectangle, every proportion is a linear relationship.

Linear relationships are sometimes referred to as a **shifted porportion**, since the graph is just like that of a proportion, but with the line shifted up to a different starting point, instead of (0,0).

In a proportional relationship, the slope is the constant of proportionality. In a linear relationship, where there is no constant of proportionality, slope is the next best thing—it still tells you how much the dependent variable changes by for every change in the independent variable. If you can imagine taking that line for Elizabeth's savings and sliding it down the graph so that it passes through (0,0), then the the slope would become a constant of porportionality.

In the graph below, two lines with different slopes are shown. One is the same line from before, Elizabeth's savings. The new line shows her brother Todd's savings.

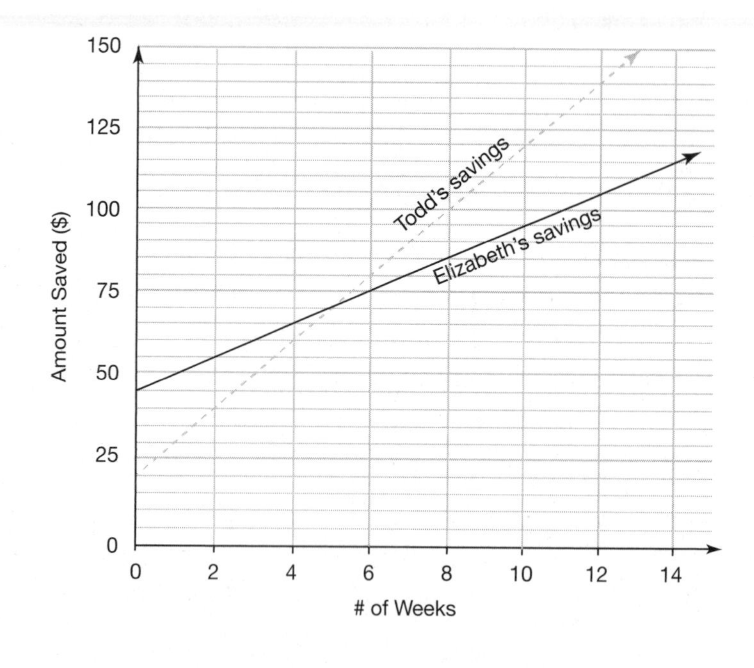

Since Todd's line is steeper than Elizabeth's, it tells us that he is saving *more* money per week. We can figure out exactly how much money per week by using the triangles:

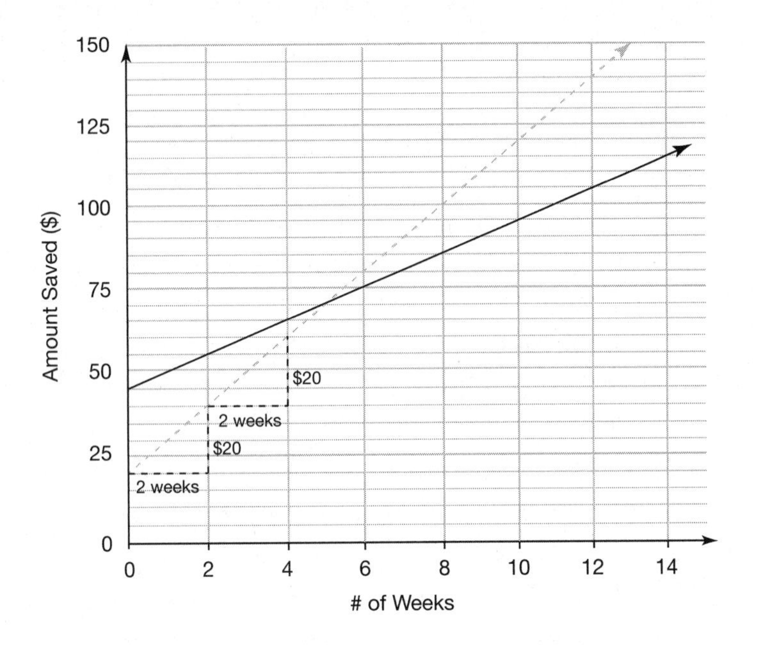

$20 in 2 weeks is a rate of $10 per week.

Formulas for Slope

Using triangles has given us a way to see what slope is. We can also think about slope in an equation form by giving names to the two legs of a slope triangle.

$$\textbf{slope} = \frac{\text{rise}}{\text{run}}$$

or

$$\textbf{slope} = \frac{\text{change in } y}{\text{change in } x}$$

Rise is the length of the vertical side of these right triangles, which is how much the *y* variable changes from one point to another—such as the amount of money earned or the distance the car has traveled. *Run* is the length of the horizontal side of these right triangles, which is how much the x variable changes from one point to another—often it is the amount of time that passed between the selected points, such as weeks for Elizabeth or hours for a road trip. **Rise is up or down. Run is left or right.**

Make sure your child remembers to divide the change in y by the change in x. Many students reverse the fraction when finding slope.

MISCONCEPTION ALERT!

Many of the misconceptions that students have about slope arise because students are used to dealing with proportional relationships, where you can find the unit rate by dividing the y-coordinate of any point by the corresponding x-coordinate. With linear graphs like the ones above this does not work because the starting value makes the relationship non-proportional.

In Elizabeth's graph, if we divide the coordinates of various points, we will see that the ratios are not equivalent. Some of the points on the line are (1,50), (6,75), and (10,95). But $50 \div 1 = 50$; $75 \div 6 = 12.5$; and $95 \div 10 = 9.5$; so the rate of change would not be constant. But we know the rate of change *is* constant because the graph is a straight line (and because we know from the verbal description that she saves the same amount each week).

Negative Slope

Sometimes graphs can have a negative slope, meaning that the y variable is *decreasing* as the x variable increases. Suppose, for instance, that we were to graph how far an airplane is from its destination at certain points during its flight:

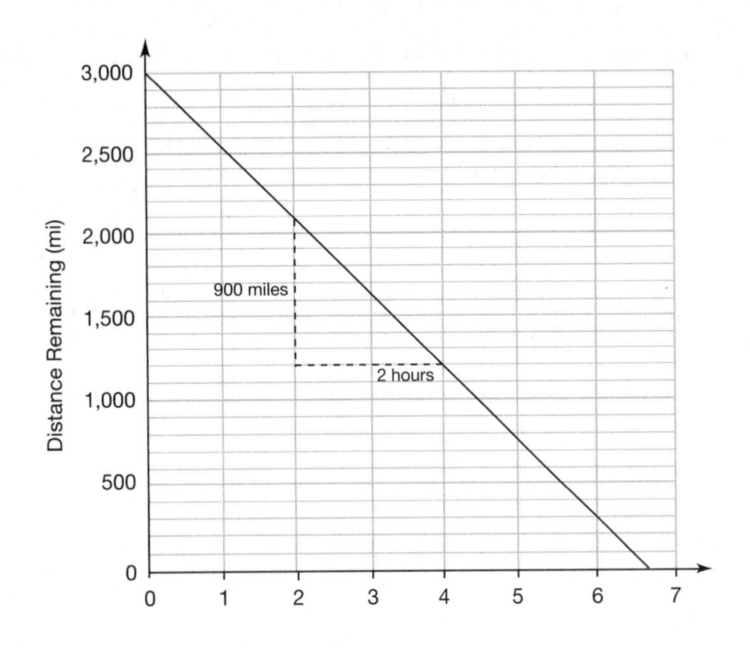

In this case, the graph line is going down as the graph moves to the right. This is considered a negative slope. When you find the slope triangles, the rise will be a negative number, since going down is the same as rising by a negative amount.

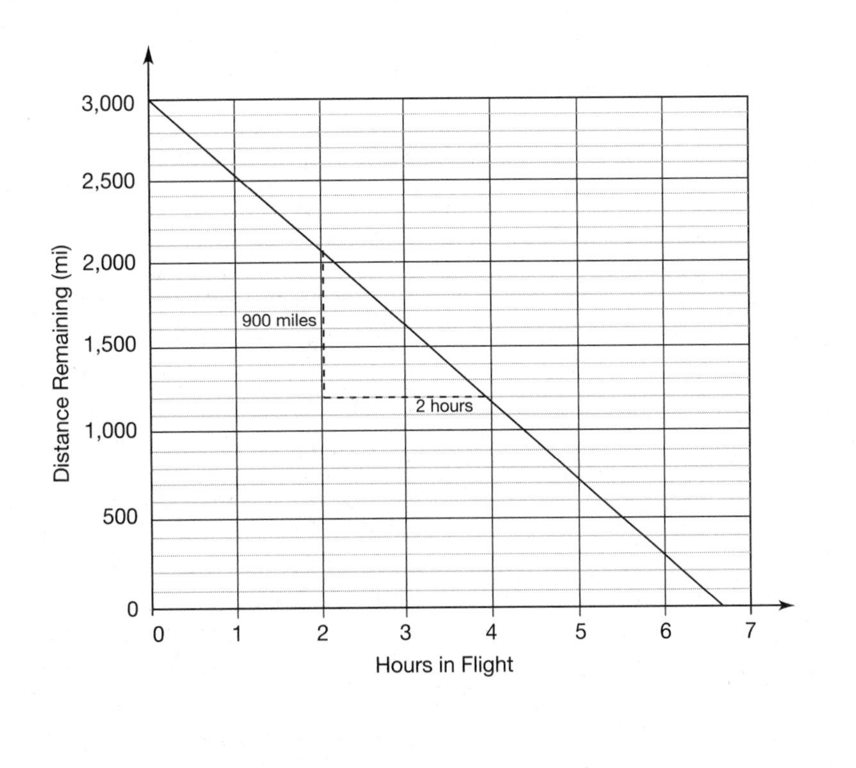

Since the airplane moves 900 miles closer to its destination every 2 hours, the slope is –450 miles per hour.

Non-Linear Relationships

Not all relationships fit a straight line when graphed. In high school, your child will study lots of different relationships that are non-linear. For now, it is enough to know that they exist, and to be able to identify when a graph is non-linear. Here are a few examples:

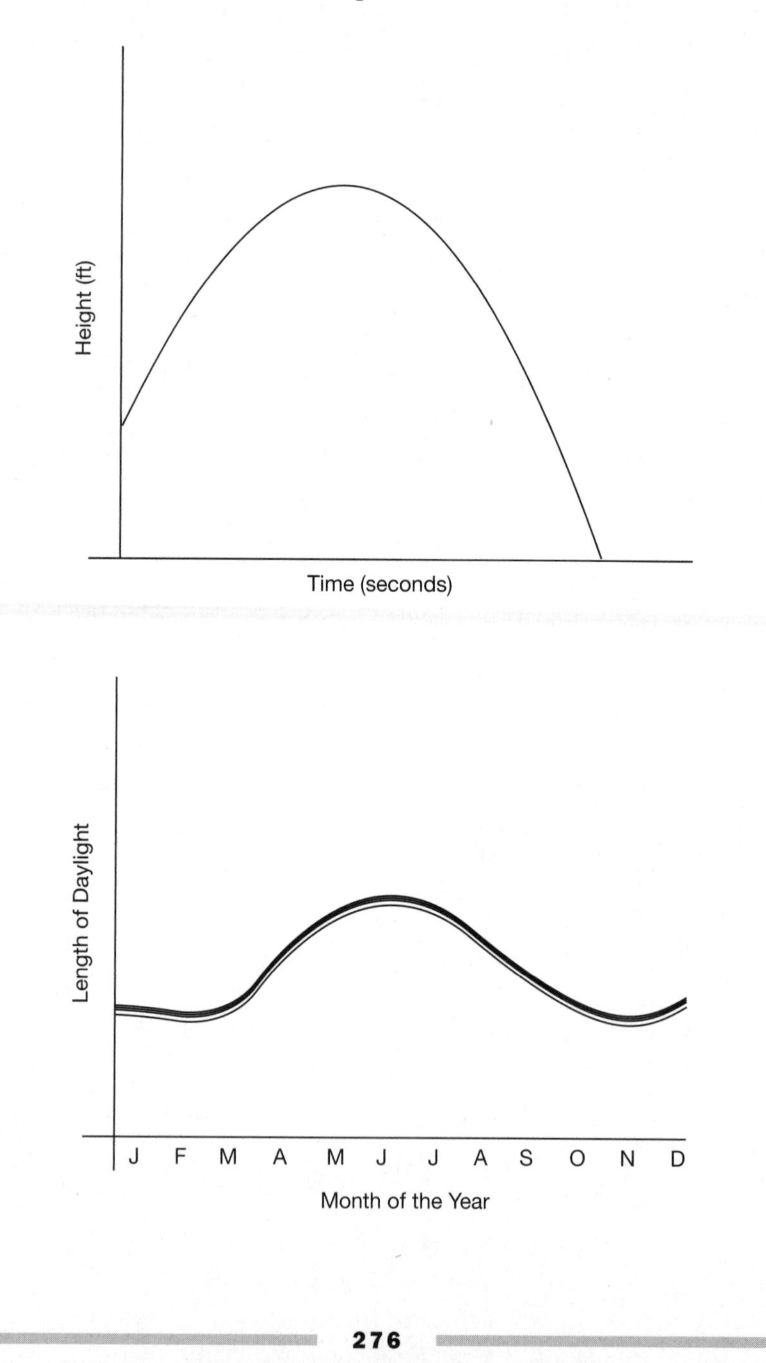

Linear Relationships in Other Representations

Just like proportional relationships, linear relationships can be represented in different forms. We have already seen a few examples of verbal descriptions and graphs. In any of the forms, your child should be able to identify both the slope and the starting value. These two numbers are extremely important because they tell you everything there is to know about a linear relationship.

Linear Relationships in Tables

Of the four representations, many students find tables to be the easiest to understand. A table is a very simple but elegant way to show a relationship between two variables—patterns immediately become apparent. The table below shows information about an online movie rental service. Each of the two variables has its own row.

Number of movies rented	0	5	10	15	20
Total cost ($)	10	18	26	34	42

In this table, there is a linear relationship between the variables. As you move across the table, both variables are increasing in consistent intervals: The number of movies is increasing by 5, and the total cost is increasing by 8. It's always important to know which variable is dependent and which is independent. Since the total cost depends on how many movies you rent, cost would be the dependent variable (y), and the number of movies would be the independent variable (x). To find the rate of change, or slope, we divide the change in cost by the change in the number of movies.

$$\text{rate of change} = \frac{\text{change in } y}{\text{change in } x} = \frac{8}{5} = 1.6$$

Now we know everything there is to know about this relationship. The starting value, 10, tells us that before you even rent any movies, there is a $10 charge. This may be a joining fee or a monthly fee. The rate of change, 1.6, tells us that every movie you rent costs an additional $1.60, since your total charge is increasing by $1.60 per movie.

Here's another example. At a carnival, the table below shows the total cost (y), depending on how many ride tickets you buy (x):

x	0	5	10	15	20	25
y	7	10	13	16	19	22

In this case, the rate of change is $\frac{3}{5}$, and the starting value is 7. The 7 means that it costs \$7 to get into the carnival, and the $\frac{3}{5}$ (or 60¢) would be the cost per ride—perhaps tickets come in packs of 5 for \$3.

It's simple to make a graph based on a table. All you have to do is plot the points with the coordinates from the table. In this case, we will graph (0,7), (5,10), (10,13), etc. Then, connect the dots!

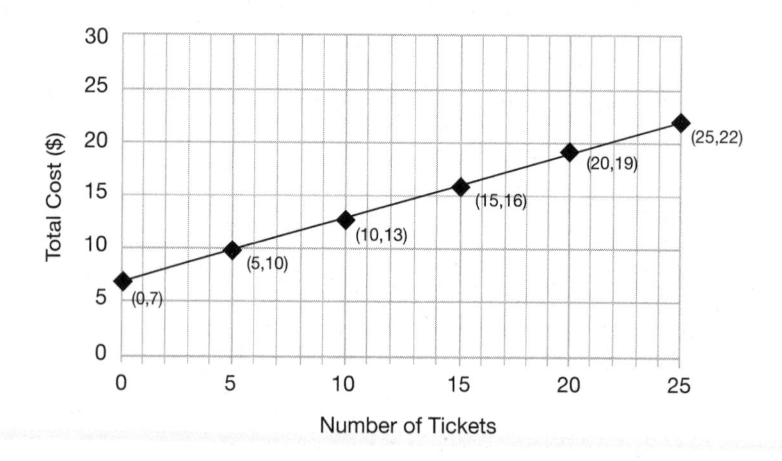

From each point to the next in this graph, we move 5 units to the right and 3 units up (this may cause your child some confusion, since we're moving up not 3 *boxes*, but 3 *units*). You can also find the slope by drawing a triangle connecting any two of the points. In this graph, we only have one point that hits the grid perfectly. But we can use any two points, since we know the coordinates:

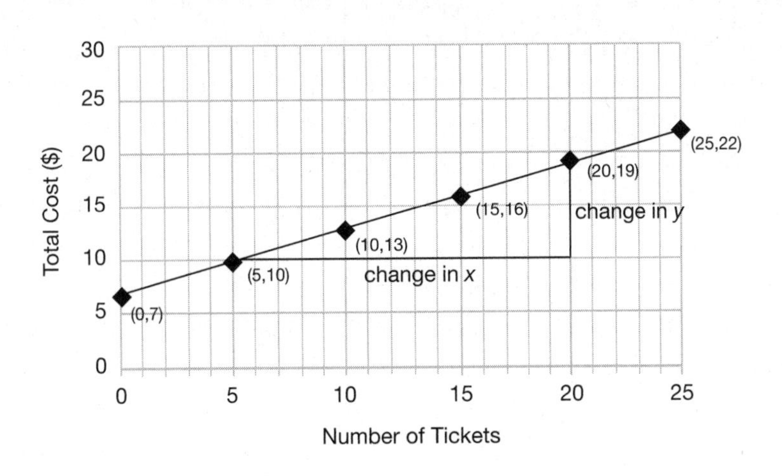

To find out how much the change in x and change in y are, all we need to do is subtract. Our right triangle has vertices at (5,10) and (20,19). So the change in y is $19 - 10 = 9$, and the change in x is $20 - 5 = 15$.

$$\text{slope} = \frac{\text{change in } y}{\text{change in } x} = \frac{19 - 10}{20 - 5} = \frac{9}{15} = 0.6$$

So the slope is $\frac{9}{15} = \frac{3}{5}$, the same rate of change we found in the table. And the 7 that we saw as the starting point in the table also shows up in the graph—it's the point where the line hits the y-axis, at (0,7).

y = mx + b

The power of algebra comes from the fact that a few symbols can be used to say a lot. Now it's time to see how equations can be written to describe linear relationships that are displayed in graphs or tables or described in words.

An Equation for a Proportion

Let's review what we covered in Chapter 4 about writing equations for proportions. Suppose Kyree earns $8 per hour for babysitting. We could make a graph to show this proportional relationship.

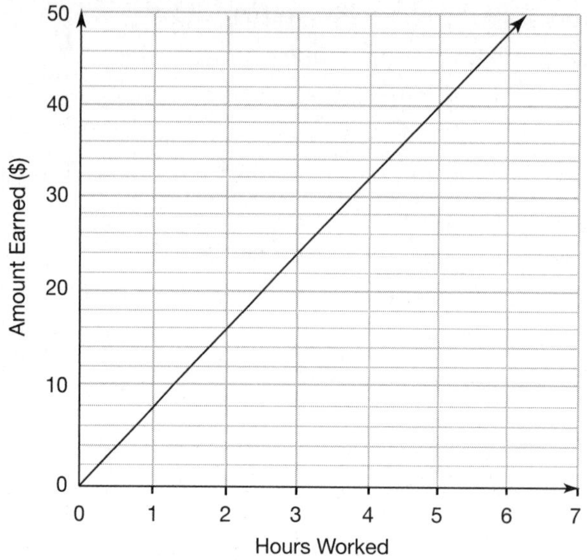

Another way to show the relationship is in a table:

Hours worked	0	1	2	3	4
$ earned	0	8	16	24	32

In the table, it is easy to see that each y value ($ earned) is 8 times the corresponding x value. If x stands for the number of hours Kyree babysits and y stands for the amount he has earned, we can find y by multiplying x times $8 per hour. So we have the following simple equation:

$$y = 8x$$

In this equation, x is the independent variable—Kyree can choose how many hours he will babysit. y is the dependent variable because the amount that he earns *depends* on how many hours he works. These two symbols say a lot—they describe the relationship in perfect detail and give instructions for how Kyree can figure out his earnings, no matter how many hours he works. Or he can use it to figure out how many hours he needs to babysit in order to meet a certain financial goal, if he has one.

To generalize this to any proportional relationship, we have used the equation $y = kx$, where x is the independent variable, y is the dependent variable, and k stands for the unit rate, slope, rate of change, or constant of proportionality—whatever you want to call it, they are all the same thing for a proportional relationship.

Generalizing to Linear Relationships

When writing an equation for a linear relationship, the process is a little more complicated. Since the starting value is not (0,0), there is an additional step and an additional term in the equation.

We can come up with a general form for a linear equation by making one slight change to the situation above. Kyree is still babysitting for $8 per hour. But let's say he is saving up to buy a computer, and he already has $122—this will be his starting point. He wants to know how much money he will have in total after babysitting for a certain number of hours.

We already know how to put this relationship into a table and a graph. In a table, we start at $122 and go up by $8 for every hour, or $40 for every 5 hours, or whatever makes sense depending on how much babysitting Kyree is planning to do. For a graph, we could start at (0,122) and follow the same slope of 8 that we had before (up 8, right 1).

The graph below shows two lines: one for the proportional relationship where Kyree earns $8 per hour, with 0 for a starting point, and one for the linear relationship where Kyree earns $8 per hour, with a starting point of $122.

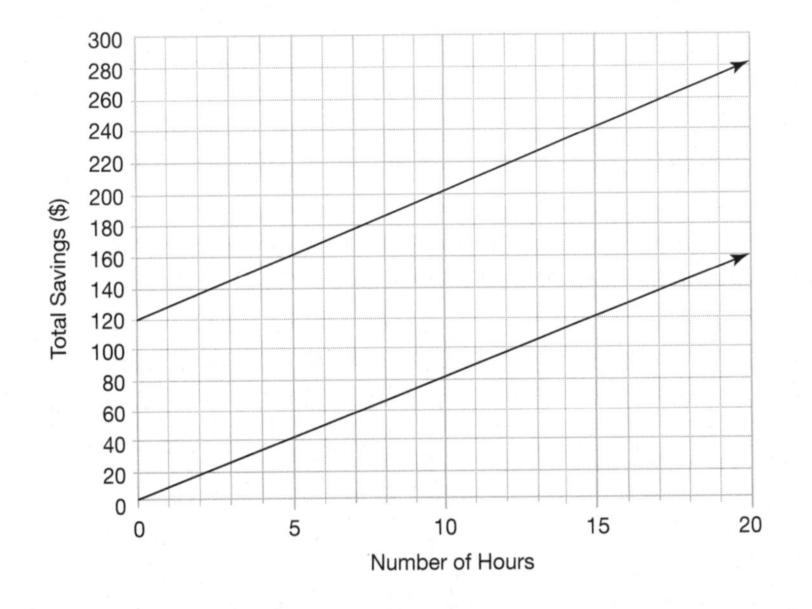

But to turn it into an equation is a bit trickier. So let's figure it out by looking at the table and comparing it to the table we had earlier:

Hours worked	0	1	2	3	4
$ earned	0	8	16	24	32
Total $	122	130	138	146	154

In this table, the bottom row shows how much Kyree has saved so far for his new computer, *after* the given number of hours of babysitting. Of course the numbers are going up by $8 each time, so we are still multiplying $8 by the number of hours. But then we are *adding* the $122 to each number to find the total, since that amount does not change, no matter how many hours Kyree has worked. We have our equation:

$$\text{Total} = \$8 \times \text{hours} + \$122$$

To put it in fancy algebra symbols:

$$y = 8x + 122$$

We can generalize that in any linear relationship, with a rate of change and a starting value, the equation will be:

$$y = (\text{rate of change}) \times x + (\text{starting value})$$

Mathematicians love to abbreviate, so we use m to stand for the rate of change, and b to stand for the starting value (don't ask why those letters!). The general equation for a linear relationship is:

$$\boldsymbol{y = mx + b}$$

x and y are the independent and dependent variables; m is the slope or rate of change; and b is the starting value.

Another name for the starting value of a linear relationship is the **y-intercept**. On any graph, the starting value occurs when $x = 0$, which means the point is on the y-axis, directly above or below (or on) the origin. Since this is the point where the line crosses the y-axis, we call it the y-intercept.

Example

Suppose a cell phone company shows customers this graph to explain their charges:

The graph has a non-zero y-intercept at (0,15). This tells you that the cell phone plan has a starting value: It might be an activation fee, a payment for the phone, or a monthly fee that is charged to your account even if you don't use any minutes. This is b. We can find m either from slope triangles or a table:

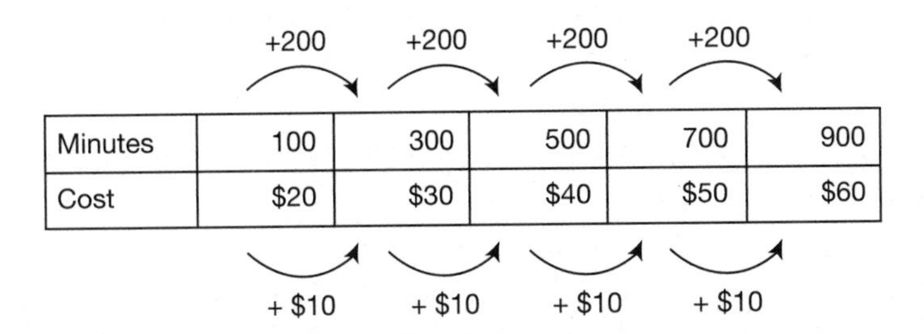

+200 +200 +200 +200

Minutes	100	300	500	700	900
Cost	\$20	\$30	\$40	\$50	\$60

+ \$10 + \$10 + \$10 + \$10

Every 200 minutes used costs an additional \$10. So the rate of change is:

$$\frac{\$10}{200 \text{ min}} = \$0.05 \text{ per min}$$

We can write an equation in the form $y = mx + b$ using this information:

$$y = 0.05x + 15$$

The equation tells us that for any number of minutes (x), you can multiply by \$0.05 and add \$15 to find the total cost for the month.

For instance, if you use 200 minutes in a month, then x is 200 minutes. $0.05 \times 200 + 15 = 25$, so it would cost \$25 for 200 minutes. If you use 850 minutes, then $x = 850$. So the total would be:

$$y = 0.05 \times 850 + 15$$

$$y = 42.5 + 15$$

$$y = \$57.50$$

You could use this equation to calculate the cost for any number of minutes.

An Abstract Example

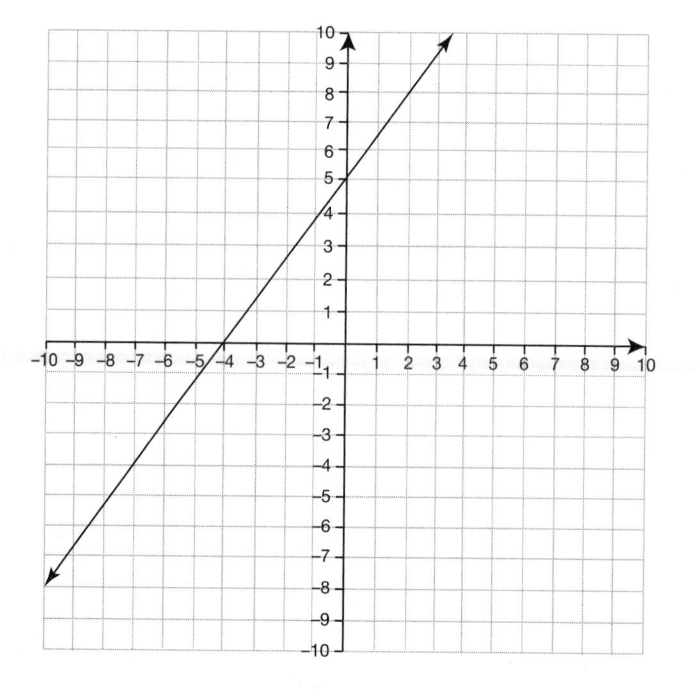

This graph has no context, as will often be the case in algebra. We can still write an equation for the line, though it will not have a specific meaning. Some of the points on the line are (–8,–10), (–4,–7), (0,–4), (4,–1), and (8,2). We can subtract coordinates as we did earlier to find the change in x and the change in y. It does not matter which coordinate pairs you choose for this, as long as you use the same pair to find both the change in x and the change in y.

$$\text{slope} = \frac{\text{change in } y}{\text{change in } x} = \frac{(-1)-(-4)}{4-0} = \frac{3}{4}$$

The slope is $\frac{3}{4}$. Of course, you could simply draw a slope triangle on the graph to find this ratio.

For b, we have to think about this graph a little differently. This line has no context, and all four quadrants are shown. We can assume the line goes on forever in both the positive and negative directions. It has nothing that we can call a starting point. So instead, we look at the y-intercept. In this graph, it is (0,–4). So b is –4, and our equation is:

$$y = \tfrac{3}{4}x - 4$$

Just as before, if you choose any value for x and substitute it into the equation, you will get the corresponding value of y. For example, when $x = 8$:

$$y = \tfrac{3}{4}(8) - 4$$

$$y = 6 - 4$$

$$y = 2$$

As we saw in the graph, (8,2) is one of the points on the line.

Notice that if you choose 0 as your x-value, the y-value will be –4, since $\frac{3}{4}(0) = 0$. In fact, in any linear relationship, when $x = 0$ then $y = b$. That's because in the equation $y = mx + b$, plugging in 0 for x cancels out the mx term, leaving only b on the right side.

Zero Slope

What happens when m is 0? For example, $y = 0x + 2$. If you choose any value of x, multiplying by 0 will give you 0. Therefore, no matter what value x has, y will always be 2. We can see this in table form:

x	0	1	2	100	3,285.9
y	2	2	2	2	2

And in graph form:

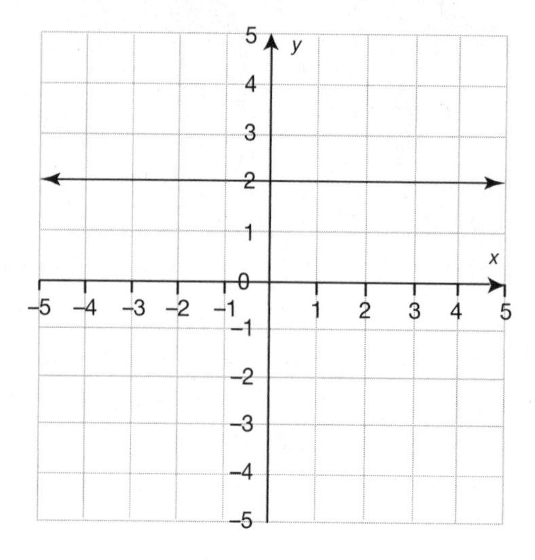

With a slope of 0, the line neither rises nor falls. The y coordinate always has the same value, so we call this a **constant relationship**. Since $0x$ is always 0, we can rewrite the equation:

$$y = 0x + 2$$
$$y = 0 + 2$$
$$y = 2$$

A line with a slope of 0 is always horizontal.

Summary

In this chapter, we have tackled the most challenging algebraic concept in all of middle school math: linear relationships. We have seen the many representations—verbal, equation, graph, and table—and have practiced converting between the different forms.

Through numerous examples, we have seen repeatedly that "math is life." There are so many ways algebra can help your child make sense of the world around him once he understands the concepts and the relationships that it describes.

In Chapter 10, we will look at another side of this idea—"life is math"—as we cover data and statistics and the ways in which they are used to tell us about our world.

Life Is Math

How to Make Sense of Data and Statistics All Around You

Data and Statistics Are Everywhere!

It is impossible to escape the barrage of data and statistics that surround us every day. With screens of all sizes, billboards, and city buses everywhere, anybody with an advertising budget is out to convince you of something—and statistics are their greatest tool. Politicians, businesspeople, corporations, educators, doctors, reporters, and sportscasters are all savvy about the way they package and present data to elicit a very specific response from their audience. As a consumer and a thinking individual, you'll benefit from interpreting these statistics intelligently and knowing when the story as presented may not be complete.

Have you ever watched sports on TV and wondered why the announcers report some of their obscure statistics? This team has won 92% of games when they were ahead by 3 or more runs in the 7th inning. This running back has gotten a first down 34% of the time on 3rd and 8 in the rain. Computer technology has made it easy to dig up this kind of information, but what does it really tell you? A good sports fan knows there's much more to the game than these numbers. Who has an injury? What kind of game has it been for the running back, and how important is this game in the big picture?

When a commercial claims that a product is 85% more effective than the competition's product, what does that mean? If a cell phone company claims that its network is 20% faster than the competition's, is this to be believed? And if it is true, does it matter? What is the downside that comes with it? Is it worth the price you might have to pay? With statistics all around us, it is easy to buy into everything we are told—and not only to believe the numbers but also to believe they are as significant as we are meant to think.

There is a famous line by the British economist Ronald Coase: "If you torture the data enough, it will confess [to anything]." Mathematicians have taken this to mean that data can be made to say anything—to prove any point you have set out to prove. If a company tests its product and the test does not go exactly as planned, there is always a way to manipulate the data until it will say what the company wants you to think. This chapter will explain some of the tools used to manipulate data, giving you a more informed perspective on statistics all around you and a starting point for helping your child with the statistics he will learn in math class.

What Are Data and Statistics?

Many people do not realize that there is a difference between data and statistics. **Data** is a set of numerical information. Suppose a survey asks 20 adults how many children they have. The data might look something like this:

1, 5, 2, 0, 0, 3, 1, 2, 1, 4, 0, 8, 2, 1, 3, 3, 1, 0, 3, 5

Statistics are ways of summarizing or describing the data. Some statistics might include the following facts:

- The average number of children is 2.25.
- The largest number of children anybody in the survey has is 8.
- 20% of the people surveyed do not have any children.
- 95% of the people surveyed have fewer than 6 children.

Statistical Questions

Some questions are considered **statistical questions** and others are not. A non-statistical question usually has a simple answer that is a single fact. Statistical questions have more complicated answers and usually involve large numbers of answers, which may vary quite a bit and can often be interpreted in a variety of ways.

The following are examples of non-statistical questions:

- What color hair do you have?
- How tall is your child?
- How many people live in your home?
- How long was the last book you read?

The following are examples of statistical questions:

- How tall are the 7th graders in your child's school?
- How many people live in each home on your block?
- How many pages are in the last ten books you have read?

As you can see, almost any question may be turned into a statistical question if it is written to include answers from large numbers of people.

Displaying Data

If you pay attention, it is easy to see graphs all over the place. You might also notice that a graph is seldom left uninterpreted. If a graph shows up in a commercial, it's a sure bet there will be a voiceover to tell you exactly what the graph means. Why is that? Because if you look too closely, you might draw your own conclusions and see something in the graph other than what the advertiser wants you to see—so they make sure to tell you what to see and think.

We will now learn about three types of data displays—dot plots, histograms, and box plots.

Dot Plots

Dot plots, sometimes called **line plots**, are one of the most straightforward ways to display a data set.

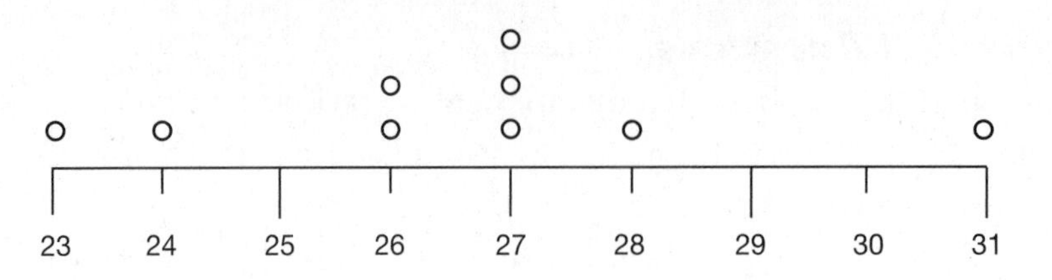

Each value in the data set is represented by a dot, or sometimes an X, that floats above a number line showing all the values in the data set. In most dot plots, every whole number—from the lowest number in the set (the minimum) to the highest (the maximum)—is shown on the number line, even if no data values fall on a particular number. When a number occurs more than once in the data set, markers are shown in a stack to represent each occurrence of that number, such as 26 and 27 in our example. The data set in the dot plot above is {23, 24, 26, 26, 27, 27, 27, 28, 31}.

Example

Suppose Mr. Garcia's 1st-period class earned the following scores on a recent math test:

84, 88, 85, 89, 90, 92, 85, 90, 88, 79, 72, 80, 90, 83, 83, 88,
85, 99, 83, 84, 76, 83, 88, 81, 88, 80, 89, 90, 91, 76, 88, 87, 100.

We can display this data on a dot plot by creating a number line that shows the full range of the data. Since the lowest score is 72 and the highest is 100, we make a number line showing every whole number from 72 to 100. Then we place a mark on top of the number line for each value. When there are several data points of the same value, the marks stack on top of each other.

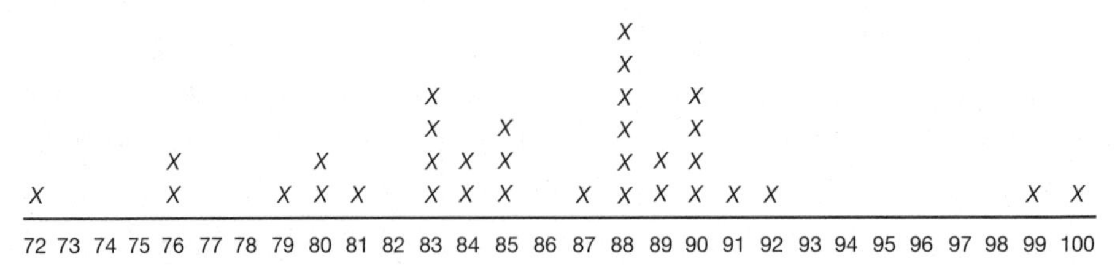

One convenient thing about making a dot plot is that in the process, we have placed the numbers in order. This will be useful for several reasons later on in the chapter.

Histograms

A **histogram** is a type of bar graph that is similar to a dot plot. Instead of including every single whole number in the range like a line plot does, a histogram divides the range into equal **intervals**, or **bins**, and shows a count of how many data values fall in each interval.

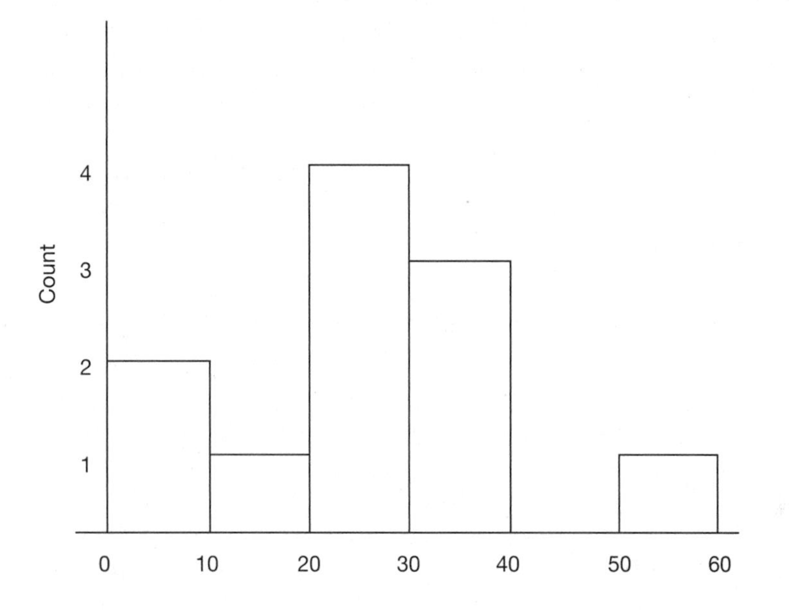

In this histogram, we can see that there are two data values between 0 and 10, one value between 10 and 20, four values between 20 and 30, three between 30 and 40, none between 40 and 50, and one between 50 and 60. A histogram makes it easy to see large amounts of data, but it does not have the precision of a dot plot. We know there are four values between 20 and 30, but we do not know what those numbers are—they could all be 20, they could all be 29, or they could all be different.

Example

Let's use the data from Mr. Garcia's 1st-period class again. The data ranges from 72 to 100. To keep the numbers clean, let's have our histogram go from 70 to 100. We can divide this range into 6 intervals, if each interval is 5 units. Let's organize the data:

Interval	Scores
70–75	72
75–80	76, 76, 79
80–85	80, 80, 81, 83, 83, 83, 83, 84, 84
85–90	85, 85, 85, 87, 88, 88, 88, 88, 88, 88, 89, 89
90–95	90, 90, 90, 90, 91, 92
95–100	99
100	100

Numbers that fall on the border between two intervals, such as 85, go into the interval to the **right** of the value. So 85 goes in the interval from 85 to 90, not 80 to 85. It really could go either way, but there has to be a consistent way of doing it. That is, we can't have some 85s in one interval and others in another, or put all the 85s *and* all the 90s in the 85 to 90 bin. Then our histograms would be misleading.

Next, we count how many values are in each interval.

Interval	Count
70–75	1
75–80	3
80–85	9
85–90	12
90–95	6
95–100	1
100	1

And we have our histogram:

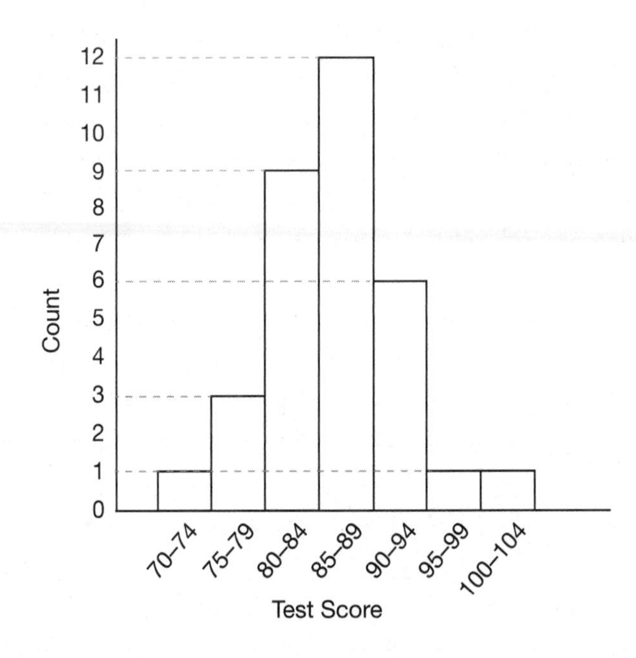

Notice that the x-axis is divided into equal-width bins and that each bin has a range of 5 numbers. Neatness counts for a lot in math. One way you can support your child in graphing is to make sure he has plenty of graph paper, and uses the grid accurately.

In terms of labels, the x-axis of a histogram should always say what the data is. The y-axis of a histogram should always say "Count," because that's what it is—a count of how many numbers fell within each interval.

NOTE

For most histograms, it is good to have between 5 and 10 bins. Fewer than that, and the histogram doesn't show much information—it just dumps all the data in big piles. More than that can be too much information.

Many students tend to want to make intervals of 10. Encourage your child to find ways to have between 5 and 10 equal intervals, even if it means having intervals that don't involve easy, round numbers.

Box-and-Whiskers Plots

A **box-and-whiskers plot**, sometimes called a **box plot**, is another data display that your child will be expected to create and interpret. A box plot divides a data set into four equal groups, called **quartiles**, and shows visually where those divisions fall. In order to divide the data, we need five important numbers, called the **five-number summary**.

- First, we'll look for the lowest and highest numbers in the data set, called the **minimum** and the **maximum**.
- Then we'll find the number in the middle of the data set—the number that divides the data into two equal halves. This is called the **median**.
- Once we have two equal halves, we will divide each half into two equal quarters. The numbers that divide the halves evenly are called the **lower quartile** and the **upper quartile**. The lower quartile is the median of the lower half of the data, and the upper quartile is the median of the upper half of the data.

I often compare the five-number summary to the fingers of a hand. If you look at the fingers of a hand, there are four gaps between them. These gaps are where the data falls, with an equal number of values in each gap.

Example

Let's use the 1st-period data again. Fortunately, our dot plot makes it easy to list the numbers in order:

72, 76, 76, 79, 80, 80, 81, 83, 83, 83, 83, 84, 84, 85, 85, 85,
87, 88, 88, 88, 88, 88, 88, 89, 89, 90, 90, 90, 90, 91, 92, 99, 100

First, identify the minimum and maximum. The minimum is the smallest number in the set, 72. The maximum is the largest number in the set, 100.

Now, to find the median we look for the middle number in the set. This data set has 33 numbers. Therefore, the 17th number in the set will be the median because it will have 16 numbers before it and 16 numbers after it.

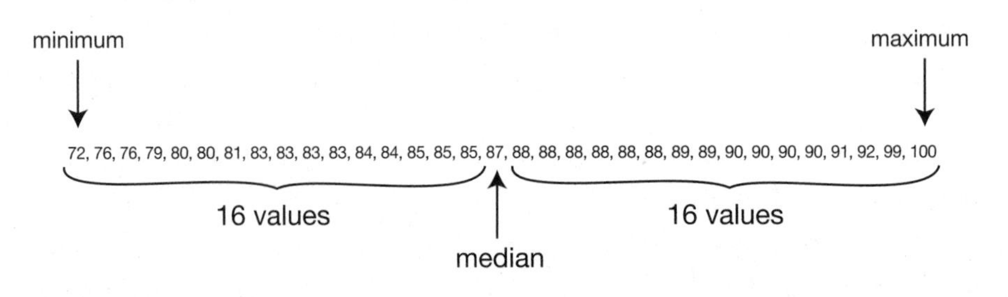

The next step is to find the lower and upper quartiles. Since the lower and upper halves each have 16 values, we are going to divide the set into equal quartiles with 8 numbers in each.

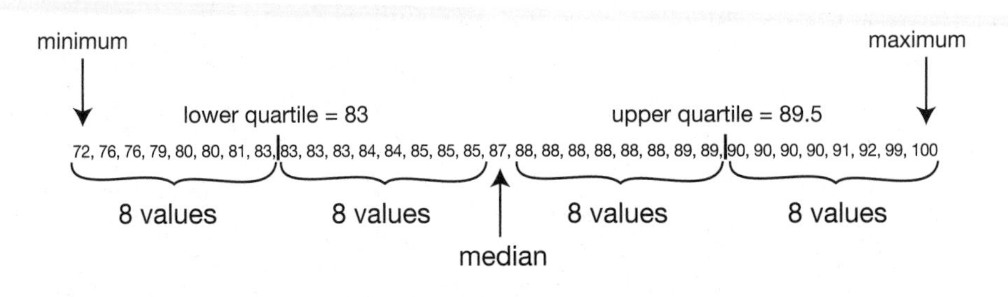

The numbers 83, 87, and 89.5 divide the set into four equal groups of data. 83 is the lower quartile, and 89.5 is the upper quartile. Later in this chapter, we will discuss how these numbers came about.

We now have our five-number summary:

Minimum	72
Lower Quartile	83
Median	87
Upper Quartile	89.5
Maximum	100

Now, let's create the graph. A box-and-whiskers plot has a silly name, but it is named that for a reason: It has a box and two whiskers. The parts of the box and the whiskers are made up of the numbers in the five-number summary.

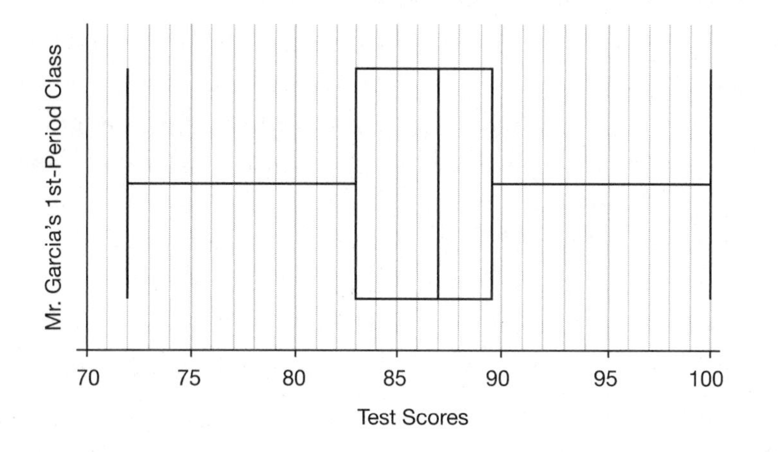

Do you see how the box plot is made of five vertical lines connected by horizontal lines that form the box and the whiskers? The vertical lines at the far ends of the whiskers are the minimum and the maximum. The left and right edges of the box are the lower and upper quartiles, and the line in the middle (but not necessarily the center) of the box is the median.

Reading and Comparing Data Displays

One of the purposes of displaying data is to be able to see two different data sets visually to make comparisons. All three data displays we have seen can be used for comparing data sets, but each has its own strengths and weaknesses. A dot plot allows you to see exact data values, which is especially useful for situations with small amounts of data and numbers that are fairly close together. Histograms are great for larger data sets, but you lose the accuracy of knowing the exact data values. A box plot has even less specific information, and is most useful for comparing the five-number summaries of two data sets.

When comparing data displays, three things that mathematicians often use are spread, shape, and center. We will use dot plots to describe these three concepts.

Reading and Comparing Dot Plots

The **spread** tells you how far apart the data is, from the minimum to the maximum. By seeing how wide a data set is spread, you can make judgments about consistency and predictability. For example, suppose you are choosing between two basketball players for your fantasy basketball team. One player has a wide spread of points per game, from 10 to 35. Another player has a smaller spread, from 19 to 27. Their dot plots may look like this:

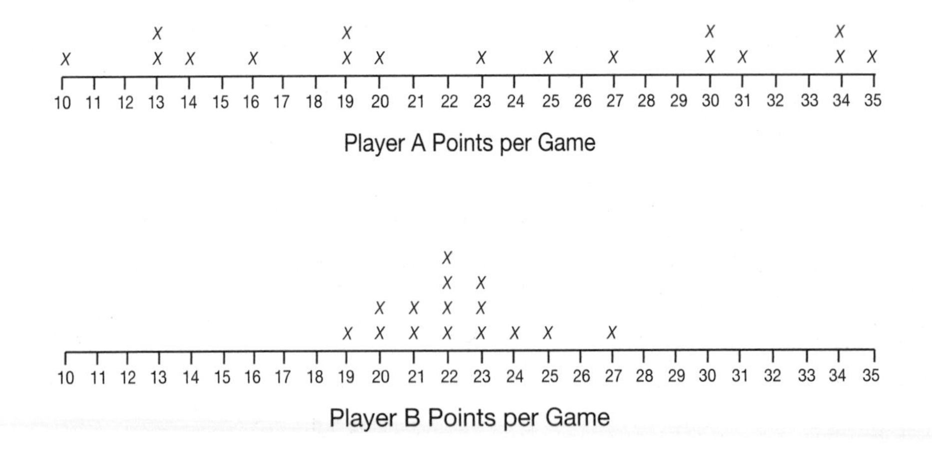

Player A Points per Game

Player B Points per Game

In this case, you need to make a judgment call. Is it better to choose the player who occasionally scores 30 or more points in a game, but often scores less than 20, or the player who consistently scores more than 20 points?

The **shape** of a graph tells you how the data values sit in relation to one another. For Player B, we can see that there is a large **cluster** of data. Player A's data, on the other hand, has several small clusters, and a few **outliers**—data values that fall far from the rest of the data.

The **center** is a third way of comparing graphs. When you look at a graph, sometimes you can make an estimate of where the middle of the data might fall. For Player B, it's pretty clear that the center is somewhere around 22 or 23. For Player A, it's a little harder to tell because of how inconsistent the data is. Later in the chapter, we will learn about different ways that mathematicians describe the center of a data set and how we can find the center of a data set more accurately.

Comparing Histograms

When you look at a histogram, you can see how far apart the data is spread and how the data is shaped. Is the data distributed evenly across a wide range, or does it cluster in one or more groups? Take, for example, the following two histograms, which compare test scores of two different math classes:

Examples

1.

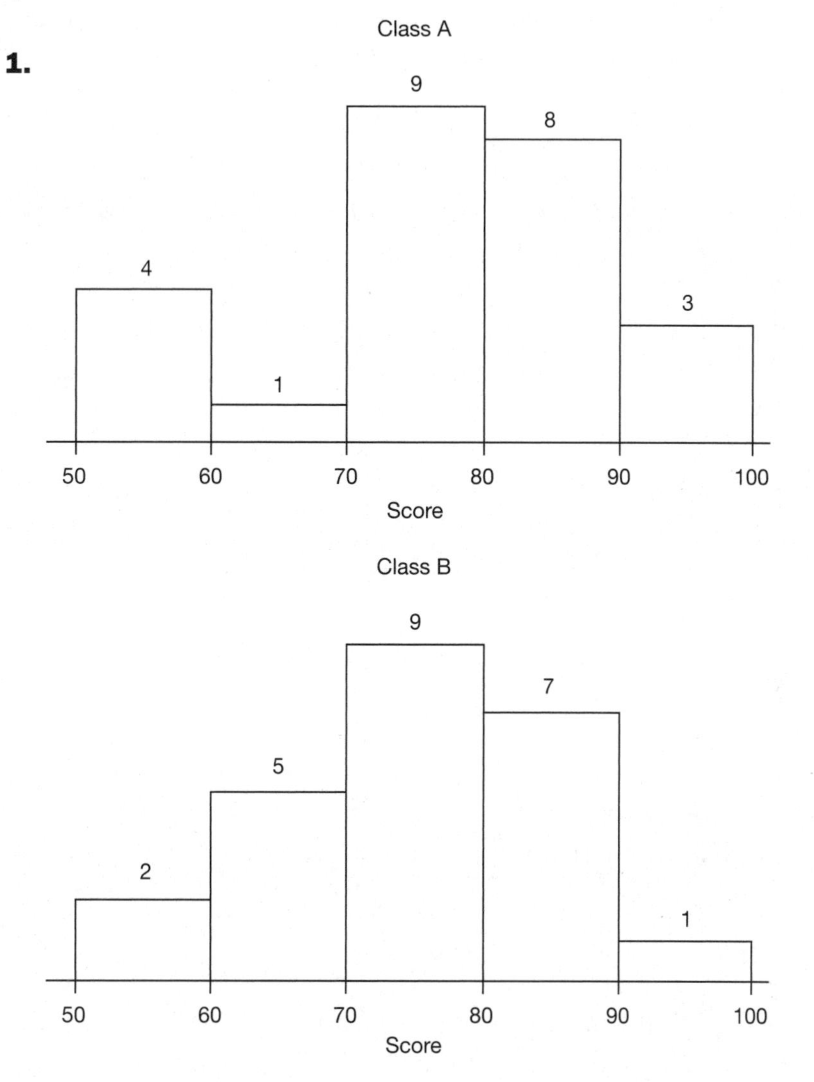

It may be difficult to say from these displays which class did *better*, but there are lots of ways we can analyze the data.

First of all, both classes seem to have pretty similar ranges—that is, both classes have students who scored in the 50s and at least one student who scored in the 90s.

Second, notice that Class B's histogram seems to have some symmetry to it. At the high and low ends, the 50s and 90s, there are small numbers of students, with more students in the middle, and the most students at the center, in the 70s. You may hear this referred to as a **bell curve** or a **normal distribution**, though that is beyond the middle school curriculum.

Class A has a very different shape. Class A has a small cluster of students in the 50s, and a larger cluster in the 70s and 80s. We could speculate on why this happened, but the important thing is describing the phenomenon that the data shows.

2. The following histograms show the values (in thousands) of homes sold in two different cities over the same time period.

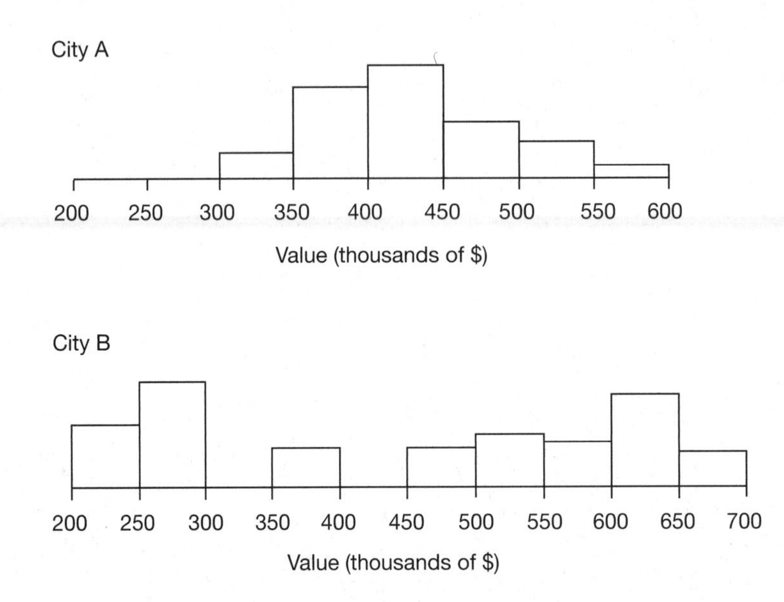

City A has a normal distribution, and all of the homes sold for between $300,000 and $600,000. By comparison, City B has a very different distribution. First, notice that it seems to have two clusters. There is a sizable group of homes that sold for lower values—$200,000 to $300,000. But most of the homes in City B are clustered between $450,000 and $700,000. Therefore, we can say that City B has a wider range of home values, or a wider spread. In City A, home prices are more consistent, and seem to stick pretty close to the $400,000 to $450,000 range.

There is a lot you could read into these comparisons. If you were a realtor trying to sell homes in these two cities, it might be in your best interest to tell your clients what you think it means. It may be less predictable what price a home will sell for in City B, or it may depend on which part of town the homes are in.

Comparing Box-and-Whiskers Plots

Box-and-whiskers plots are great tools for comparing different data sets and have some clear advantages over histograms, under certain circumstances. Suppose you are comparing two data sets that are very different in size. One set has about 100 values in it. The other has close to 10,000. Even if the data values are all in the same general range, a histogram would not be a great choice because the histogram with the larger data set would have so much data that the other histogram would look very small, no matter how the data differs. Of course, one could choose to use a histogram for that very reason—the graphs would be very misleading, which could be advantageous for proving a certain point or justifying an argument.

Since a box-and-whiskers plot only uses the five-number summary, it is great for comparing different size data sets.

Example

A school did a survey of students and teachers to find out who sends the most text messages in a month. 300 students and 25 teachers were surveyed. The data is displayed below.

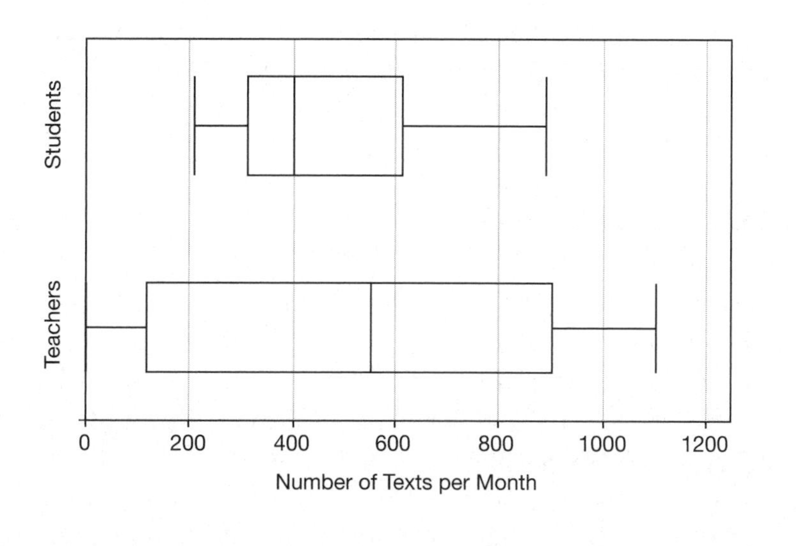

There is a lot of information that can be derived from this simple display.

Let's examine the teachers' box plot first. Teachers ranged from 0 texts to around 1,100 texts in the month. The lower quartile for teachers is around 150, so we know that 25% of teachers sent less than 150 texts in that month, and similarly, 25% of teachers sent more than 900 texts. The median for teachers was around 550 texts.

Students had a minimum that was over 200—no student in the survey sent (or admitted to sending) fewer than 200 texts in the month. The maximum for students is somewhere between 800 and 900, and the median is 400.

These statistics give us lots of ways to compare the two groups. Teachers' texting data has a wider range—from 0 to 1,100 rather than 200 to 900. That means that while some teachers do not text at all, and at least 25% of teachers text less than *any* student, the teachers who text a lot text a *lot*—at least 25% of teachers sent more texts than any student. Students are more predictable. The box for students' data is much narrower than teachers, telling us that 50% of students sent between 300 and about 625 texts. A typical student does not seem to vary much from the median of 400 texts—in fact, 25% of students sent between 300 and 400 texts. That's a lot of data in a very small range. On a histogram, it would appear as a large cluster.

DATA & STATISTICS VOCABULARY

One thing you may have noticed by now is that there is a ton of vocabulary in statistics. You can support your child's learning by making sure she knows the meanings of the vocabulary words. To a student who does not understand the terminology, math can be very daunting. Look back through the topics in this chapter and ask your child about the **bolded** vocabulary terms.

Measures of Center

When large amounts of data are gathered, it is convenient to describe the data with just a few numbers. As we learned, the five-number summary is one way of condensing data to a more useful form. Now we will build on that idea to explore several other numbers, called **measures of center and variability**, which tell us things about a data set.

- Measures of center include the **mean**, **median**, and **mode** of a data set. They are numbers that describe where the center of a data set is in different ways.
- Measures of variability are ways of describing how predictable or consistent a data set is. For this we will use the range, interquartile range, and mean absolute deviation.

Statistics is about number crunching—taking large sets of numbers and boiling them down to just a few numbers that reveal something about the data. In the real world, mathematicians, statisticians, scientists, and businesspeople rarely do this work by hand because their data sets are usually very large. There are plenty of computer programs designed to handle these things quickly. But to be able to use statistics in a meaningful way, students must first develop a sense of what the statistics are all about, and that happens when the student knows how those numbers are calculated. In middle school math, these routine calculations are done multiple times, to the point of tedium, but it is with that purpose in mind.

THE IMPORTANCE OF SHOWING WORK

As you will soon see, the calculations involved in statistics can be somewhat daunting. For your child, this is a great opportunity to learn strategies for keeping numbers organized when doing long calculations. When calculating means, and eventually mean average deviations, help your child slow down and show each step in the calculations. It will pay off later on, when your child is facing complex calculations involving large numbers of data and algebraic symbols. Baby steps are critical to building understanding.

Finding the Median

As we briefly covered when discussing box-and-whiskers plots, the **median** is a number that divides the data set into two equal halves. That is, half the data is greater than the median value, and half is less.

If there are an odd number of values in the data set, then the median will be the middle number when the data is put in order. If there are an even number of values in the data set, then the median will be the number halfway between the middle two numbers when the data is put in order.

An Odd Example

Let's find the median of the data set 8, 12, 19, 6, 9, 14, 3.

First, we need to rearrange the numbers in order: 3, 6, 8, 9, 12, 14, 19.

Now find the middle number, which is 9. The median is 9 because there are three values less than 9 (3, 6, and 8), and three values greater than 9 (12, 14, and 19). With the exception of the median itself, half the data is larger than the median, and half the data is smaller than the median.

An Even Example

Now let's find the median of the data set 17, 24, 12, 8, 16, 10, 8, 15.

First, put the numbers in order: 8, 8, 10, 12, 15, 16, 17, 24.

Since there are an even number of values, there will be two numbers in the middle—12 and 15. So our median is the number halfway between 12 and 15. There are a few different ways to find this middle number. One way is by dividing the difference between 12 and 15 in half:

$$15 - 12 = 3$$

$$3 \div 2 = 1.5$$

$$12 + 1.5 = 13.5$$

Another way is to find the **average** of 15 and 12 by adding them up and dividing by 2:

$$15 + 12 = 27$$

$$27 \div 2 = 13.5$$

You could practice your algebra skills by proving that these two methods are equivalent for any two numbers.

A third way is to use a number line and find the midpoint:

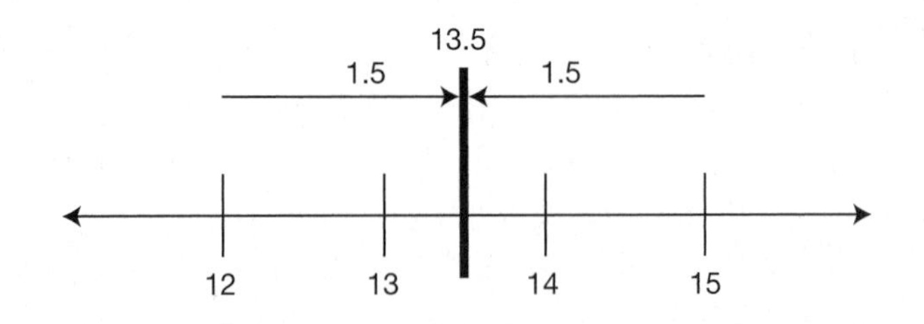

You can easily see from this number line that 13.5 is the same distance from 12 and 15.

13.5 is the median of our data set. Half the values in the set (8, 8, 10, 12) are smaller than 13.5, and half the values in the data set (15, 16, 17, 24) are larger than 13.5.

MISCONCEPTION ALERT!

One mistake that many students make is forgetting to put the numbers in order when finding the median. Since the order of the numbers as given is often arbitrary (that is, the original list could be in any order without changing the data set), it will not give you a reliable median. Check your child's work on median problems to make sure he is putting the numbers in order.

Sometimes the median may be a number that is repeated in the data set. For example, consider the set 15, 23, 28, 28, 28, 32, 44, 91. The numbers are already in order, and there are 8 total numbers. The 4th number and 5th number are both 28, so the median is 28 (28 is halfway between 28 and 28).

Finding the Mean

The **mean** is the average of a data set. To find the mean, add up all the values in the data set and then divide by the total number of values in the data set.

Examples

1. Suppose your child's test scores are 89, 96, 83, 77, and 84. Find the average of these test scores.

   First add up the test scores: $89 + 96 + 83 + 77 + 84 = 429$

   Now divide by the total number of values in the data set. There are 5 values in the data set:

   $429 \div 5 = 85.8$.

   Notice that even though the test scores are all whole numbers, the mean is a decimal. This happens most of the time, since the final step in finding the mean is division.

2. Find the average of these bowling scores: 125, 136, 115, 55, 112, 124.

First add them up: $125 + 136 + 115 + 55 + 112 + 124 = 667$

Now divide: $667 \div 6 = 111.1\overline{6}$

VISUALIZE: MEAN AS AN EQUAL SHARE

Your child's teacher may present the mean as an equal share. Suppose, for example, that five friends spent the following amounts of money on supplies for a party: $9, $17, $13, $22, and $20. If the friends want to redistribute cash so that each person spends the same amount, the person who spent $9 might start by giving some money to the person who spent $22. The person who spent $13 could give a few dollars to the person who spent $22, and so on, until everybody has the same amount. When they finally get to a point where everybody has spent the same amount of money, that amount will be the average of the original numbers.

When your child is studying the mean in school, try solving one or two of her homework problems in this way. Use objects (checkers, pennies, grains of rice, whatever) to make a pile for each value in the data set, and redistribute them until everything is equal. In the end, you may have to "cut" (i.e., imagine cutting) the last few objects into smaller pieces, but you will end up with the average.

Finding the Mode

The **mode** of a data set is the most common number in the set. Sometimes a data set may have more than one mode, if there is a tie for the most common number.

Examples

1. The mode of 17, 13, 18, 15, 17 is 17 because it occurs more often than any other number.

2. In the dot plot below, the mode is 25. You can easily see that there are more 25s in the data set than any other number.

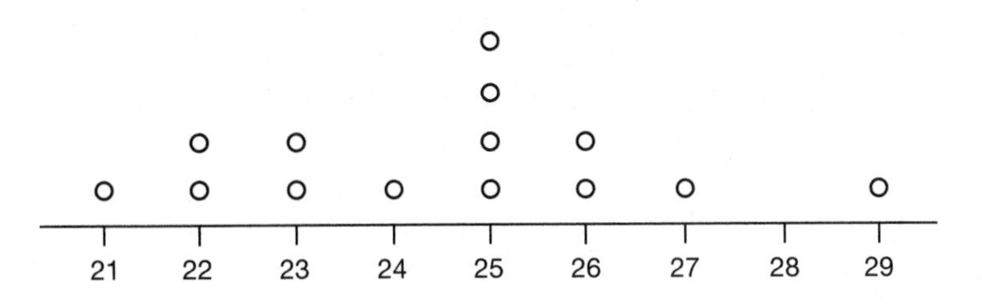

3. In the next dot plot, there are two modes, 96 and 100.

```
                X                             X
          X     X                       X     X
    X     X     X     X     X     X     X     X
  --+-----+-----+-----+-----+-----+-----+-----+--
   94    95    96    97    98    99   100   101
```

4. Some sets have no mode. No numbers repeat in the set 12, 15, 81, 3, 26, 0. This set has no mode.

Relevance of the Measures of Center

We hear about averages all the time. Most of the time, **average** refers to the mean—like a batting average, a GPA, or the average rainfall for a certain month. The median is not used as often, but in fact, it is sometimes a more accurate representation of a data set.

Let's look again at the two examples we used for finding the mean. In the first example, the test scores, we had a mean of 85.8. This is a pretty good representation of the student's test scores. It is very close to three of the numbers in the data set (83, 84, and 89), and it's close to the middle between the highest and lowest values (77 and 96). For this reason, it makes sense to think of 85.8 as the "center" of the data. It is a good representation of a *typical* value in the set.

In the second example, the bowling scores, the mean is not a very good representation of the data. The mean was $111.1\overline{6}$. But 5 out of 6 of the scores were above the mean, and only two of the numbers (112 and 115) were even close to it. The reason this mean is so inaccurate is because of the 55. This is an **outlier**, a number that is far away from the other numbers in the data set. This bowler may have had a bad day, or maybe it was the person's first time bowling, or perhaps he was letting a child take his turns. Whatever the story, the outlier has a big impact on the average for the data set.

Instead of the mean, let's look at the median for the bowling scores. In order, the scores are: 55, 112, 115, 124, 125, and 136. There are 6 numbers, so we look between the 3rd and 4th to find the median.

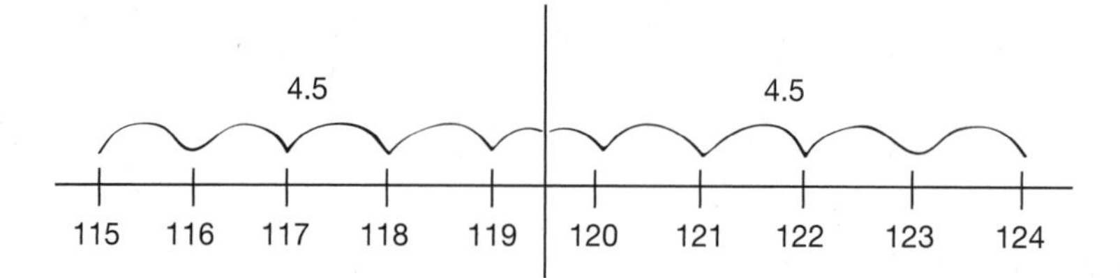

So the median is 119.5. Even though this isn't too far from the mean (only about 8 points), it is a much more accurate representation of the data as a whole. 119.5 is within a few points of most of the numbers in the set: 112, 115, 124, and 125.

The mean can change significantly if outliers are included in a set. This is why a 0 on a missed homework or test can be so devastating to a student's average in a class. But the median is *not* affected by outliers. In the bowling example, that 55 could have been 112 or 90 or 0, and it wouldn't change the median at all. Likewise, the 136 could have been a 300 without changing the median.

For this reason, median is often used with data sets that are likely to have outliers—for example, home prices in an area or salaries for a company are often described by their median, rather than their mean. If a company pays 95% of its employees salaries between $30,000 and $80,000, while a few executives make over a million dollars, the mean might end up somewhere in the hundreds of thousands. The median will be somewhere below $80,000, and is therefore a much more accurate representation of salaries at that company.

Measures of Variability

In middle school, your child will learn three different numbers that can be used to describe how much **variability** a data set has. Variability is a funny mathematical term because it is difficult to define. It is a way of describing how wide and wild the data in a set is.

For example, the set 8, 9, 6, 10, 9, 6 does not have much variability because the numbers are all close together, and some of them even repeat. If you had to guess the next number to come up in this set, there's a good chance you wouldn't be far off. A set like 10, 98, 862, 4, –394, 17 has a lot of variability because the numbers are all over the place—they are unpredictable.

Range

The simplest measure of variability is **range**. The range of a data set is how far apart the numbers are, from smallest to largest. To find the range, just subtract the minimum of the data set from the maximum.

In a box-and-whiskers plot, the range is the distance between the ends of the two whiskers.

Examples

In the data set 27, 21, 25, 16, 35, the range is 19, the difference between the largest and smallest values: 35 – 16 = 19.

In the data set 56, 37, 99, 84, 104, 77, the range is 104 – 37 = 67. This indicates that this data set is more variable than the previous example because it has a larger range.

Interquartile Range

Earlier in this chapter, we studied the five-number summary, which includes the minimum, lower quartile, median, upper quartile, and maximum of a data set. While range is the difference between the minimum and the maximum, the **interquartile range** is the difference between the upper and lower quartiles. In a box-and-whiskers plot, the interquartile range (sometimes abbreviated IQR) is the width of the box. It tells you how spread out the **middle 50%** of the data is.

Example

Consider the data set 9, 18, 15, 22, 16, 9, 22, 27, 10.

The five-number summary is:

Minimum:	9
Lower Quartile:	9.5
Median:	16
Upper Quartile:	22
Maximum:	27

In this data set, the range is 18, which is 27 – 9. The IQR is 12.5, which is 22 – 9.5.

FORMULAS TO REMEMBER

Range = Maximum – Minimum

Interquartile Range = Upper Quartile – Lower Quartile

Mean Absolute Deviation

If you have any experience with statistical studies, you may be familiar with the term **standard deviation**. Standard deviation is a measure that is often used to show how far the data values are from the mean of the data set. The formula for calculating standard deviation is too complicated for the middle school level, but mean absolute deviation works as a simplified substitute.

The **mean absolute deviation** of a data set is the average distance of all the values in a data set from the mean. That's a mouthful. In other words, you find out how far each value is from the mean and take the average of these differences. Let's think of it in steps:

1. Find the mean of the data set.
2. Subtract the mean from each data value.
3. Take the absolute value of each of these differences (i.e., make them all positive).
4. Find the mean (average) of all these differences.

Let's demonstrate with a simple example:

A Simple Example

Consider the small data set 26, 28, 42. In order to find the mean absolute deviation (which we can abbreviate MAD, though math has nothing to do with negative emotions), we first need to find the mean:

$$\text{Mean} = (26 + 28 + 42) \div 3 = 32$$

Now we need to find each number's distance from the mean, or **deviation**.

$$26 - 32 = -6$$

$$28 - 32 = -4$$

$$42 - 32 = 10$$

Notice that for 26 and 28, the deviations are negative because the values are less than the mean. The next step is to take care of that by making everything positive—that's why it's called the mean *absolute* deviation, because we're using the absolute value. We do this because we do not want values above and below the mean to cancel each other out—we want every value to count. If you're curious, try following the process *without* taking the absolute value—just average all those positives and negatives together. You may be surprised at what happens.

Next we take the absolute deviations and find their average. $(6 + 4 + 10) \div 3 = 6.\overline{6}$. So the MAD for this data set is $6.\overline{6}$.

A More Complicated Example

Let's find the MAD for 16, 24, 18, 11, 3, 18, 18, 26.

The mean is 16.75: $(16 + 24 + 18 + 11 + 3 + 18 + 18 + 26) \div 8 = 16.75$.

The absolute deviations are 0.75, 7.25, 1.25, 5.75, 13.75, 1.25, 1.25, and 9.25. Find these by simply subtracting 16.75 from each of the numbers in the data set, and then writing the absolute value of each one.

So the mean absolute deviation is the average of these numbers:

$$\begin{aligned}&(0.75 + 7.25 + 1.25 + 5.75 + 13.75 + 1.25 + 1.25 + 9.25) \div 8\\&= 40.5 \div 8\\&= 5.0625\end{aligned}$$

So on average, the numbers in the data set 16, 24, 18, 11, 3, 18, 18, 26 are 5.0625 away from the mean.

Comparing Data Sets

Now that we know how to calculate the measures of center and variability for a data set, we can do some analysis by comparing different data sets.

If you like sports, you have certainly heard many comparisons of measures of center. Averages are used all the time in sports. Medians are less common but also very useful.

It's easy enough to compare two baseball players based on their batting averages, or two basketball players based on their average points per game. If one player's average is 25.2 points per game and another player's average is 18.9 points per game, it may seem obvious which player scores more. But sports fans rarely limit their comparisons to this level of simplicity. How good is each player's field goal percentage? What about their 3-point averages? Which player is more agile? Can he break through the opponents' defenses?

All of these questions make comparisons more difficult but also more interesting. If a player's worth could be measured by one single number, there would be very little to discuss.

Choosing and Comparing Measures of Center

More often than not, measures of center make it pretty simple to compare two things. If you're looking at a large set of data to compare two athletes, usually the athlete with the higher mean will also have a higher median, and the comparison is simple. Mode is less important, since there may only be one data value that repeats.

Occasionally, the mean and median will tell different stories. Suppose you are comparing two basketball players in terms of their points per game.

Player A	Player B
Mean = 22.9 Median = 24	Mean = 25.3 Median = 18.5

In this case, Player B has a higher mean than Player A, but Player A has the higher median. How can we compare these players?

If Player A's median is 24, it means that in half of his games, he has scored more than 24 points. By comparison, Player B scored *below* 18.5 in half his games. A fan could reasonably say that Player A is more likely to score higher in a game than Player B, even though he has the lower mean.

Why is Player A's mean lower? As we saw earlier, outliers can have a big impact on the mean of a data set. In general, when the mean and median are far apart, it has to do with outliers. It might be one outlier that is really far from the rest of the data, or a handful of outliers that aren't quite as far out, but definitely don't fall in with the rest of the pack.

Based on the statistics we know, we can assume that Player A is pretty consistent, reliably scoring in the low 20s. Player B probably had some high outliers—a small number of really great games—to boost his average.

Which player is better? It can be argued either way, as sports fans and commentators often do. It will often come down to a question of what the team needs in terms of speed, chemistry, leadership, defense, etc.

Comparing Measures of Variability

The table below shows statistics for home sale prices in two different cities (numbers are in thousands of dollars).

City A	City B
Mean = 352	Mean = 218
Median = 205.5	Median = 220
Mode = 125, 599	Mode = 215
Range = 814	Range = 41
IQR = 24	IQR = 27.5
MAD = 241.4	MAD = 12.9

You can tell from the measures of center that City B is more consistent. The mean, median, and mode are all close together, and the measures of variability confirm it—home prices in City B don't vary much. In City A, however, things are different. With the mean and median so far apart, there must be some high-priced outliers.

To have such a wide range, City A must have both very expensive and some relatively inexpensive homes. City B's small range shows that all of its prices are very consistent (at least for the given data set). The mean average deviation for City A reveals that the average ended up being pretty far from most of the data. This is because of the expensive outliers, which raised the average well above most of the home prices. By comparison, City B has a small mean average deviation, reinforcing the idea that the home prices are all fairly close to the mean, and therefore all fairly close together.

Interestingly, City B has a larger IQR than City A. City A's large outliers do not seem to affect the IQR in a big way. This is because the IQR measures how spread out the *middle 50%* of data values are. Since the outliers do not fall in the middle 50%, they do not affect the IQR. Therefore, we can assume that most of City A's homes fall in a pretty narrow range of values, with the exception of the outliers—the very expensive homes. If we could see a histogram, the cluster for City A in the low 200s would be narrower than City B's cluster.

Measures of center and variability give us a lot of ways to compare two different data sets, but they require us to read a bit into the numbers—if you don't, someone will always be happy to furnish you with their opinion of what the statistics mean. The more you practice calculating these statistics from data sets, the better you will understand their meaning and therefore be able to compare them in clever ways. Practice with your child by comparing statistics about something she is interested in, whether it's sports, money, or any other topic. You can always find interesting statistics on the Internet.

Sampling from Populations

Have you ever seen statistics in the news and wondered how anyone could possibly know that kind of information? For example, when the media reports the President's approval rating? If they say 54% of Americans approve of the job the President is doing, how could they know that? Did they ask every American for his or her opinion? Did they ask you?

It would be nearly impossible (and very expensive) to ask every American what he or she thought of the President's performance. Yet the reports are supposed to be somewhat accurate. The pollsters and statisticians that provide the data use a method called **sampling**, in which a small portion of the whole population is selected to represent the population in a particular poll.

What Is Sampling?

Sampling is a way of using porportions to make inferences, or guesses, about a population based on data from a smaller number of individuals, called a **sample**. The word **population** refers to any whole group of individuals we may be interested in, such as the population of your state or all the people who bought cars last year.

Examples

1. Suppose there are 800 students in your child's school. Your child wants to conduct a survey to find out what activities are most popular among the students. It would be impractical to ask all 800 students, so your child might choose a sample—a smaller number of people that can easily be surveyed.

2. During elections, news media are constantly trying to predict the outcomes of elections by using statistics. Since they can't possibly ask every voter, they use sampling—they ask small groups of voters and use those people's responses to predict how the whole population will vote.

In order to get results that truly represent the population under study, it is important to choose a sample that is **unbiased**. This doesn't mean that the individuals in the sample are unbiased, but that the sample will include and represent *all* biases in proportion to the population. This will be made clearer by the following examples.

Types of Sampling

There are many different ways one might choose a sample for a survey. As we describe the following three sampling methods, we will continue with the example of your child surveying students at school about their favorite activities.

Voluntary Sampling

Suppose your child made an announcement in the cafeteria at lunchtime calling for any students who wanted to participate in the survey. Students who wished to volunteer could then approach your child to take the survey. This is an example of **voluntary sampling** because it relies on participants to volunteer. A voluntary sample does not always give the most reliable results because it can be inherently biased. Suppose none of the students who like sports volunteer because they were already outside on the court or the field when your child made the announcement. Then your child's sample would not accurately represent the whole population.

Convenience Sampling

Convenience sampling is just what it sounds like—convenient. In the case of your child's survey, the convenient thing to do might be to ask her homeroom class. Like voluntary sampling, convenience sampling often does not produce a representative sample. In some schools, for instance, homerooms may be grouped based on academic classes. If your child's homeroom consists mostly of honors students, or does not include any music students, then the sample may not accurately show how many students in the school prefer reading or music as an activity. Therefore, convenience samples are often biased.

Random Sampling

In **random sampling**, every member of the population has an equal chance of being selected for the survey. For your child's survey, this may mean asking the school administrator for a list of all 800 students, cutting all 800 names out of the printout, mixing them in a bag, and pulling 40 names without looking.

This method is considered the most reliable and unbiased because it has the best chance of removing the conditions that make convenience and voluntary sampling biased. Every student in the school is equally likely to have his opinion registered, no matter what that opinion may be. Since we are not selecting students on the basis of anything that might be relevant to the subject of the survey, we have a good chance of having every activity students like named. If half the students in the school like sports, then there is a good chance that about half the students in the sample of 40 will say they like sports.

Using a Sample to Make Inferences about a Population

When a sample is selected for a statistical study or survey, one assumes that that sample represents the whole population. To make predictions about the population, we can use proportions to scale the results from the sample up to the size of the population. This is how pollsters in the news media make predictions about who will win an election, or make judgments about an elected official's approval rating. If 62% of the people surveyed think the mayor is doing a good job, then we can assume that around 62% of the city's population also think the mayor is doing a good job.

Companies use this kind of reasoning when doing market research. If there is no major news story in a week, then a weekly news magazine might poll a focus group to find out which topic to use as the cover story to sell the most copies of next week's issue. If a large percentage of the sample says they would buy a magazine with a cover story about a certain topic, then the publisher will consider going with that story.

Example

Let's continue with our earlier example. Suppose your child asks 40 schoolmates their favorite activities, and gets the following results:

Activity	Number of Votes
Reading	12
Games	9
Music	3
Sports	16

Your child could use these data to make predictions about the whole population of the school (which is 800 students), using simple proportions. Let's use reading as an example:

$$\frac{12}{40} = \frac{?}{800}$$

Since 800 is 20 times larger than 40, we can solve this proportion by multiplying 12×20, which is 240. Therefore, we can infer that *about* 240 students in the whole school prefer reading. Your child might also be curious about what percentage of the population prefers reading. That is a simple calculation: $\frac{12}{40} = \frac{30}{100}$, so we would expect that about 30% of the 800 students prefer reading.

Summary

As your child progresses through middle school, she will learn more and more about data and statistics each year, deepening her understanding and opening herself up to increasingly sophisticated analysis. This is a great opportunity to engage in conversations with your child about the information that is presented all around us and to think critically about what is genuine and what is not.

When a report talks about an average, or makes a statement about a percentage of the population, your child will learn to ask questions about whether the sample was random and unbiased, or whether the average includes outliers that might skew it. When you see a graph, look at it closely together to investigate whether it truly shows the point that someone is trying to make with it. Does it really match what he or she is trying to say? Is there something in the data that is not being mentioned, perhaps for a specific reason?

Final Thoughts

We've been through a lot together in these pages, and no doubt you and your child have, too. I hope you have learned something valuable about math and that you and your child have had opportunities to see math in a different light. Remember to always keep it positive, ask great questions, and make connections between what you're learning and your life. I wish you all the best for success in middle school, high school, and beyond!

Acknowledgments

There are so many people who have helped me complete this project.

I'd like to thank my wife, Rachel, and my sons, Jacob and Benjamin, who tolerated me disappearing to coffee shops on so many weekends and evenings to finish this work, as well as all the family and friends who helped with childcare. I love you guys. A special shout-out to my mom for that business with the calculus.

Speaking of coffee shops, the vast majority of this book was written at the greatest little coffee shop in the world, A Muddy Cup, in Seattle. Sarah and Mat (whom I think of as "The Muddy Couple") have been good friends and great neighbors throughout the process, and I miss them terribly since I've moved away. I also owe thanks to three coffee shops in and around Philadelphia: Volo Coffeehouse in Manayunk, Saxby's in Ardmore, and Town Hall Coffee in Merion Station.

I couldn't have done any of this work without the influence of my greatest mentor and dear friend, Gini Stimpson. She taught me everything important that I know about teaching, learning, and math. And she babysat quite a few times while I was writing.

I owe an enormous debt of gratitude to all of my past students and their families, who inspired this work. I learned many of the strategies presented in this book while working at Madison Middle School in West Seattle.

Special thanks to the parents who gave their time for interviews: Mike McComb and Jen Jonson. And extra special thanks to the Nelson family for so much positive feedback and so many (so, so many) Pi Day cookies over the years.

Two Madison families stand out among the crowd: Maddy Emerson and her mom, Jen Hey, and Katie DuLong and her parents, Anne and Jeff DuLong. Maddy and Katie were 8th-grade students of mine who went along with my crazy idea to take Algebra 1 in the summer. Their families went through a lot that summer, and tolerated me the whole time. Their experience was what gave this book life.

Thanks to Jasmine Riach, one of my great Madison colleagues, who gave me the idea for the famous fruit punch in Chapter 4.

Of course there's my editor, Sheryl Posnick, who made this book happen, shaped it, and legitimized me as a writer. Thank you for your kind attention to all my double-spaces after periods and unstacked fractions!

Finally and above all I'd like to thank you, the reader, for everything you are doing to help your child. Your child deserves a great math education, and you are absolutely a part of that. Since your child may never say *thank you* for your efforts, I'm saying it now.